Nuclear Magnetic Resonance Imaging in Medicine

Nuclear Magnetic Resonance Imaging in Medicine

Edited by

Leon Kaufman, Ph.D.
Lawrence E. Crooks, Ph.D.
Alexander R. Margulis, M.D.

Foreword by

Paul C. Lauterbur, Ph.D.

IGAKU-SHOIN New York • Tokyo

Published and distributed by

IGAKU-SHOIN Ltd.,
5-24-3 Hongo, Bunkyo-ku, Tokyo

IGAKU-SHOIN Medical Publishers, Inc.,
1140 Avenue of the Americas, New York, N.Y. 10036

ISBN: 0-89640-057-3
Library of Congress Catalog Card Number: 81-82043

Fourth printing, 1983

Printed and bound in the United States of America

Contributors

William Bradley, M.D.
University of California, San Francisco
Radiologic Imaging Laboratory
400 Grandview Drive
South San Francisco, California 94080

Thomas F. Budinger, M.D., Ph.D.
Donner Laboratory and
Lawrence Berkeley Laboratory
University of California, Berkeley
Berkeley, California 94720

Lawrence E. Crooks, Ph.D.
Assistant Professor of Electrical Engineering
University of California, San Francisco
Radiologic Imaging Laboratory
400 Grandview Drive
South San Francisco, California 94080

Peter L. Davis, M.D.
University of California, San Francisco
Radiologic Imaging Laboratory
400 Grandview Drive
South San Francisco, California 94080

James M. S. Hutchison, B.Sc., Ph.D.
Department of Biomedical Physics and Bioengineering
University of Aberdeen
Aberdeen, United Kingdom

Leon Kaufman, Ph.D.
Professor of Physics
University of California, San Francisco
Radiologic Imaging Laboratory
400 Grandview Drive
South San Francisco, California 94080

David M. Kramer, Ph.D.
Department of Chemistry
State University of New York at Stony Brook
Stony Brook, New York 11794

Paul C. Lauterbur, Ph.D.
Professor of Chemistry and
Research Professor of Radiology
Department of Chemistry
State University of New York at Stony Brook
Stony Brook, New York 11794

Alexander R. Margulis, M.D.
Professor and Chairman
Department of Radiology
University of California, San Francisco
San Francisco, California 94143

Derek Shaw, Ph.D.
Oxford Research Systems
Ferry Hinksey Road
Oxford OX2 ODT, England

Jerome R. Singer, Ph.D.
Professor of Electrical Engineering
Electronic Research Laboratories
University of California, Berkeley
Berkeley, California 94720

Francis W. Smith, M.B.Ch.B., D.M.R.D., F.F.R.
Department of Nuclear Medicine
Aberdeen Royal Infirmary
Aberdeen, United Kingdom

Heather Tosteson, M.F.A.
University of California, San Francisco
Radiologic Imaging Laboratory
400 Grandview Drive
South San Francisco, California 94080

Foreword

To peer within the human body, examining internal organs with the same ease and delicate discrimination with which we can look at the skin, and with the same freedom from hazard, has not until recently been even a dream. To see the internal equivalents of a flush, a discoloration, a slight puffiness or a quiver has seemed obviously impossible, although radiologists long ago learned to discern many subtle details in their x-ray shadows. Everything is different now; new techniques, born of scientific cleverness and digital computers, have stimulated the physicians' dreams and are rapidly changing the procedures, the ambitions and the economics of diagnostic medicine. X-ray computed tomography, ultrasonic scanning and emission imaging with radionuclides have had the greatest impact, but the revolution is broader and deeper than the introduction of one or a few novel techniques. There is growing appreciation that general imaging concepts and methods can be applied to any interaction of waves, particles or fields with the body, and that each new possibility not only brings new opportunities and new potential hazards, but also shifts our perceptions of the complementary roles for which new and old techniques are best suited.

Nuclear magnetic resonance imaging is rapidly approaching a time of widespread clinical trials. It may ultimately give images as clear as those obtained with transmission computed tomography, with a potential for defining tissue chemistry, directly or by inference, in ways reminiscent of emission imaging, and may even compete with ultrasound in some measurements of blood flow and heart motion. Perhaps the technique can be as free from hazards as is visual observation; at least no ionizing radiation is involved. There are drawbacks and complexities, of course. The nuclear spins give weak signals, and the images can be obtained only with large, complex and expensive apparatus that is still rapidly evolving. It is too early to predict the forms and capabilities of the various specialized NMR devices that will emerge as technical possibilities and useful medical applications are more clearly realized. Similar observations can be made about all new diagnostic techniques, of course, and nothing more is needed to justify this volume of introductions and reviews.

But, in addition, there are the many remarkable features of the NMR phenomenon itself, and of the imaging techniques based upon it, that make books like this, the others that will no doubt follow it, and basic texts on NMR essential reading for anyone interested in the development of diagnostic instrumentation. In NMR experiments, atomic nuclei unmistakably reveal their elemental and even isotopic identities, and their molecular locations and the motions of the molecules can often be deduced as well. The imaging principle used is itself novel; there are no rays of particles or of waves, but instead there is an organized, partly coherent motion of nuclear spins within the entire volume of the region being imaged. This fundamental difference, which has led to some use of the term zeugmatography to distinguish these techniques from all others, imposes limitations and affords possibilities unique to this way of making images. Easy analogies must be avoided; resolution, contrast, sensitivity, speed and specificity are not always what they might seem, and their interrelations are not yet fully understood. Nevertheless, it is already clear that NMR techniques will be an important component of the emerging diagnostic imaging technology that will observe the interior of the body as gently and clearly as we now view its surface, but with even greater specificity and sensitivity to physiologic changes. The chapters that follow present a summary of the basic physics, examples of some current techniques, instrumentation, and images, introductions to in vivo elemental and chemical analysis and a discussion of the biological effects that may be anticipated. It is a rich and stimulating collection for anyone interested in diagnostic techniques and instrumentation.

Paul C. Lauterbur

Preface

This book has been compiled to provide practicing radiologists who have not been exposed to nuclear magnetic resonance (NMR) imaging with a succinct general introduction to the field. To aid clinicians in their evaluation of the medical implications of this rapidly developing imaging technique, we have included brief explanations of the basic physics of NMR and of the technical requirements of the modality, a descriptive survey of existing NMR imaging techniques, examples of images currently being produced by NMR imaging systems accompanied by discussions of their medical implications, as well as discussions of future medical applications of NMR imaging and an evaluation of the potential hazards of the technique.

Certainly, the objection can be made that a book on NMR imaging in medicine is premature. Only a few human-scale systems are now technologically advanced enough to begin tentative clinical studies. However, the potential of the technique has generated so much interest among the general medical public that we think a simplified survey of the major theoretic and technologic considerations of this substantially different imaging method will already prove helpful to clinicians, who ultimately will decide what systems to purchase and their most efficient diagnostic applications. It is obvious that any book on a rapidly expanding scientific field becomes outdated in the time between the completion of manuscripts and their publication. Nevertheless, the basic approaches, the basic background, and the discussion of medical implications and future applications will still be of value. By providing physicians with the basic information necessary to understand the strengths, limitations, and implications of NMR imaging as a diagnostic technique and as a technology, we hope this book will help to create an informed consumer group.

The editors are grateful for the patience and diligence with which the authors prepared their contributions and apologize in advance for all the fine research efforts that could not be included because of time and space limitations.

Mary Camp and Julie Brown were instrumental in preparing correspondence and manuscripts in a timely manner. We thank Douglas Symes and Patricia Christen for their careful and prompt preparation of most of the illustrations.

Major credit for this book rests with Ms. Heather Tosteson. As associate editor, she coordinated the material, kept authors on schedule, actively participated in the writing of several chapters, rewrote major parts of others, and edited the entire manuscript so that it would remain pertinent to the concerns of its intended audience.

May 1981

Leon Kaufman
Lawrence E. Crooks
Alexander R. Margulis

Contents

1

Potential Medical Applications of NMR Imaging

Alexander R. Margulis, M.D.

Not since the introduction of computed tomography (CT) has there been so much excitement about a new modality in radiology as there has been about nuclear magnetic resonance (NMR) imaging. Many universities in the United States, Europe, and Japan are actively performing research in this area, and major x-ray CT manufacturers throughout the world are engaged in extensive investigative programs to develop a clinical model that can operate well in the hospital environment.

Meetings held to discuss the current status of NMR imaging attract hundreds of attendees from medical, engineering, and industrial communities as well as ubiquitous government observers.

Interest in NMR derives from both its research applications and its potential usefulness as a diagnostic technique. The latter has provided the impetus for a research investment by industry that is advancing the state of the art at a pace that quickly makes any predictions of capabilities obsolete.

So much activity with limited availability of workable clinical units is a testimonial to the unique qualities that the modality possesses:

1. NMR imaging does not use ionizing radiation and so far appears harmless, except possibly for limited patient populations (Chapter 10).
2. Images of hydrogen obtained by NMR are of excellent quality, with spatial resolution appearing almost comparable to that of CT (Chapters 5 and 6).
3. NMR has the capability to generate images in response to intrinsic tissue parameters, the T_1 and T_2 relaxation times, as described elsewhere in this monograph. These parameters represent gross chemical characteristics, such as state of hydration, fat content, and presence of

paramagnetic compounds. Relaxation times provide a major degree of the tissue sensitivity of NMR imaging (Chapters 2, 3, and 5).

4. It may someday be possible to image other elements, such as phosphorus, sodium, and nitrogen (Chapter 9). Because these elements are not nearly as abundant in the body as hydrogen, and are not as sensitive to NMR, it is likely that their images will be of value only when superimposed on a map of hydrogen distribution.
5. Because it takes a finite time to register the signal emitted by hydrogen nuclei, fast-moving nuclei may not be recorded; that is, the technique is sensitive to flow. Blood circulating through a major vessel in a healthy person appears with low intensity in cross section because of the flow, but in an occluded vessel or after death, it will appear with a high intensity (Chapter 5).

By using hydrogen distribution and the distribution of its relaxation times, by recording the speed of transit of hydrogen nuclei through the plane of imaging, and by superimposing the distribution of other elements and the distribution of their relaxation times on the hydrogen map, a wealth of information will become available. However, before diagnostic application, both the chemical and the diagnostic implications of this information must be understood by physicians and researchers. Because NMR does not provide a straightforward visual representation of a single physical parameter (e.g., electron density or radionuclide distribution), a full understanding of the phenomenon requires that we become familiar with other physical concepts. This consideration is particularly important: because of the complex interdependence between any imaging technique and the image it produces, we must understand the physical basis of the NMR phenomenon and of the various methods by which the NMR response can be elicited to evaluate realistically the sensitivity of all imaging systems and imaging techniques. Not only do physicians need to become acquainted with the state of the art, but further research also needs to be done. Only in the last decade have investigators begun to explore the applications of NMR to tissue studies. Most research has been done with NMR high-resolution spectroscopy. However, attempts at predicting image characteristics by in vitro analysis of tissue samples have had ambiguous results. The resonance parameters must be more fully understood before they can have genuine biochemical importance. We are presented with the prospect of vast amounts of new data, but until the data are assimilated into a coherent intellectual frame, they cannot be realistically evaluated or applied.

NMR IMAGES

As we have noted, the NMR intensity image depends on the hydrogen density distribution of the body, the T_1 and T_2 relaxation parameters, and the speed through thc volume being imaged. In the intensity image, when using a gray scale on which white indicates high intensity, we find that fat is white, liver light gray, soft tissues gray, muscle a darker gray, bone dark gray or black, and air black because of the virtual absence of hydrogen (Fig. 1.1). With some imaging techniques, blood vessels during life appear almost black. After death, they appear light gray (Chapter 5). Thus, the image intensity of blood depends on its rate of flow. For other techniques, blood is light in the live. Many tumors appear lighter than adjacent soft tissues because their T_1 or T_2, relaxation time (or both) is long. Differentiation of white and gray matter in the brain is impressive when its relaxation parameters are taken into account.

By varying the times between signal excitations or signal observations, different tissues can be enhanced as a result of their T_1 and T_2 values. In this sequence of images of a tumor in a rat, one can see how manipulation of the T_1 parameter dramatically serves to heighten intrinsic object contrast (Fig. 1.2).

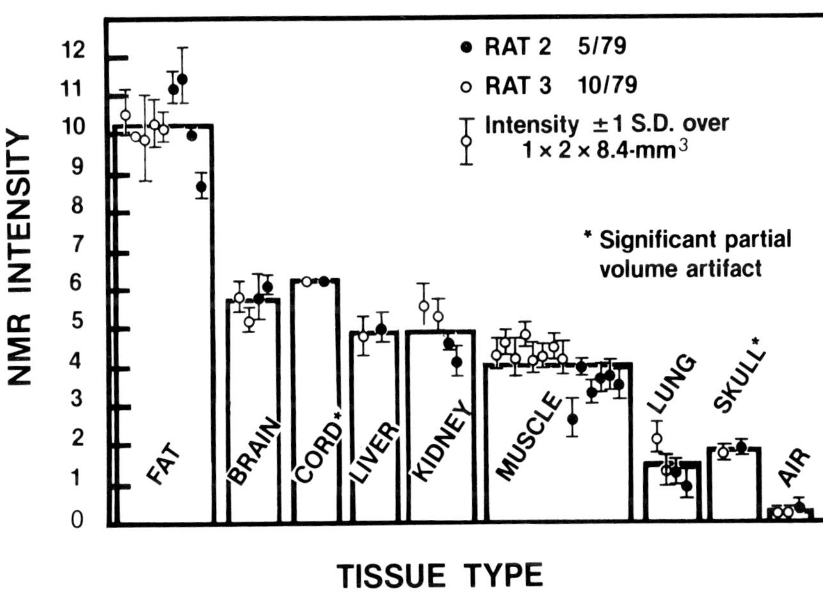

Figure 1.1. NMR intensity, or gray scale, for various tissues obtained under one set of operating conditions.

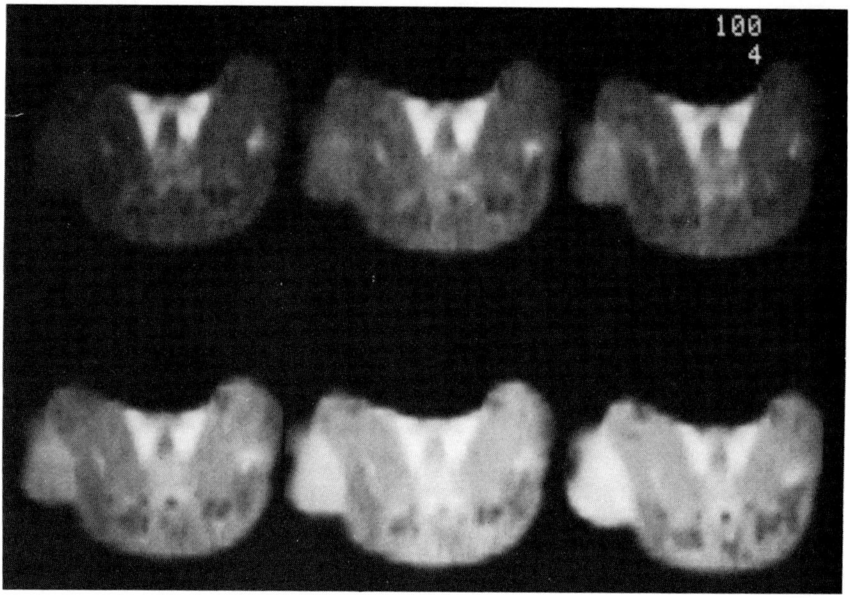

Figure 1.2. The relative and absolute gray-scale changes as operating conditions change. In this set of six NMR images of the flank of a rat with an implanted sarcoma, the T_1 parameter of the instrument changes from 0.25 to 1.5 sec from left to right, top to bottom. Note that with increasing T_1 parameter, the tumor becomes more intense with respect to surrounding muscle.

By algebraic manipulation, one can produce T_1, T_2, and hydrogen density images. Although intensity images from different systems are not comparable because of the difference in imaging techniques (due to the relative importance of the T_1 or T_2 parameter in a particular imaging process), derived images of T_1, T_2, or hydrogen density should be qualitatively comparable if obtained from systems operating at the same magnetic field strength. T_1 and T_2 values of the same tissue in different persons may vary due to state of hydration, fat content and, possibly, aging. In addition, oxygenation changes relaxation times. Systematic categorizations of T_1 and T_2 relaxation times in tissues in NMR imaging have only just begun. A full understanding of these categorizations is necessary if the technique is to assume its full potential. The observation that relaxation times are generally longer for tumors has provoked a great deal of debate about whether NMR imaging may prove particularly useful in the diagnosis of malignant tumors. Unfortunately, abcesses and hematomas also show long relaxation times (Chapter 5). It seems that, whether these parameters prove chemically important, they do serve a valuable observable contrast function.

COMPARISON TO OTHER RADIOLOGIC DIAGNOSTIC MODALITIES

Radiography

When conventional radiographs are combined with contrast media, particularly those administered by invasive methods like angiography, they offer much information. Interventional radiography, intraluminal contrast studies of the gastrointestinal tube, excretory urography, arteriography, cardioangiography, and lymphangiography are procedures that make diagnostic radiology an informative and rewarding speciality. The information thus obtained has exquisite spatial resolution. However, it is presented in shadow form, with all planes superimposed on each other. There is, therefore, the need to obtain images at 90-degree angles to gather three-dimensional information. The amount of radiation administered by these methods is sizeable.

Computed Tomography

Computed tomography is a variation of conventional radiography. The efficacy of detectors and the computer enhancement of irradiated thin slices permits the acquisition of images with superior density discrimination and true tomography. Reformatting in any desired plane avoids mistakes by presenting images in a cross section of numerous planes. Although the amount of information in these planes does not change, the perception of anatomy is easier and diagnostic mistakes are avoided. X-ray CT exposes patients to appreciable amounts of radiation and also depends heavily on the use of contrast media. Intraluminal gastrointestinal contrast media have no potential for harm, but they are not reliable in distending the small bowel. Moreover, the spatial resolution of the method does not compete with conventional single- or double-contrast barium studies. Intravenous injections of water-soluble iodinated contrast media, either by drip or bolus, have an inherent danger of iodine reactions and are not sufficiently specific in labeling. Computed tomography is incapable of distinguishing infiltrating neoplasms from inflammatory disease or even from anatomic variants. Neoplasms in the liver, pancreas, kidneys, or the gastrointestinal tract, for instance, often can be detected only if they change the outline of the organ.

Ultrasound

Sector scanning affords superb resolution and information that in some ways exceed and often complement that of x-ray CT. It is highly doubtful whether tissue signature information will ever be achieved. Nevertheless, ultrasound has the advantages of not being dependent on ionizing radiation and of being safe. The difficulty with ultrasound is the abundance of noise in the system, interference by gas and bone, and the lack of suitable contrast media.

Nuclear Medicine

Nuclear medicine offers information of high metabolic organ and disease specificity. Radiopharmacology offers the hope of an unlimited number of contrast media by inventive tagging of metabolites. The disadvantages are poor spatial resolution, lack of signals from distal planes of thick organs, dependence on ionizing radiation, and precautions required for handling radioactive material and its wastes.

Disadvantages of NMR

NMR could form a bridge among all radiologic modalities. It avoids most of their disadvantages but has some drawbacks. Having already discussed its promising aspects, I shall review the negative aspects of the modality:

1. A relatively slow imaging time, especially when imaging large objects.
2. When compared with ultrasound, conventional radiography, or even x-ray CT, spatial resolution is presently poorer, although better than that of nuclear medicine.
3. The side effects of a strong magnetic field preclude the introduction of magnetic materials into the vicinity of the machine.
4. There is the danger of harming patients who have cardiac pacemakers.
5. There is a possible danger, not yet totally explored, of local overheating of tissues when large metallic implants (for instance, prostheses) are present. Surgical clips do not appreciably heat surrounding tissues.
6. The cost of helium needed for superconducting magnets is high, and reliable refrigerators will be required.

7. The system, particularly the magnet with its high field uniformity, is difficult to manufacture, so only a limited number of laboratories are building them at this time.
8. The cost of production is currently high because customized construction is required.

Contrast Media

A short comparison of the degree of soft-tissue contrast available with the different modalities is helpful because it shows clearly the particular promise of NMR.

So far, few new contrast media have been developed for CT. In general, contrast media for CT are diluted compounds used in conventional radiography. Research in the development of contrast media for ultrasound has not been as fruitful as had been expected. Except for bubbles in tap water, which are sometimes used to identify the stomach, no other ultrasound contrast medium is now in standard clinical use. Radioactive tracers used in nuclear medicine present hazards and, at best, are burdensome to handle.

It is worth exploring whether NMR will offer the opportunity of using totally harmless contrast media, such as increased hydration, increased oxygenation, milkshakes, or similar foodstuffs for use in the gastrointestinal tract, or insufflation of air per rectum. The unique promise of NMR, however, is the possibility of achieving decisive soft-tissue contrast without having to use any contrast media, simply by manipulating the intrinsic parameters of the NMR imaging process.

MEDICAL GOALS

It behooves the radiologic community to take an orderly approach to developing and introducing NMR imaging. A slow, controlled, well-phased program will maximize returns and avoid the backlash reaction that has hampered the use of x-ray CT during the last few years. The promise of NMR should not blind us to its actual value and its actual limitations.

With the unprecedented industrial investment in research to develop NMR imaging equipment and techniques, it appears that funding agencies can concentrate scarce resources on the study of the physiologic basis and capabilities of this technology. As previously mentioned, the biochemical implications of the resonance parameters are poorly understood.

NMR may in the future offer a method for identifying various types of disease processes. It is expected that small neoplasms will be discoverable, either without the use of contrast media or with the use of totally harmless contrast media. It is hoped that in the future, when NMR imaging is based not only on hydrogen but also on the signals from nuclei of other elements, various types of neoplasms can be differentiated. It is even conceivable that by the use of the T_1 and T_2 relaxation times of the various imaged elements, the phases of growth of these neoplasms could also be recognized. But the first goal of any imaging technique, of course, must be to differentiate between healthy and abnormal tissues. Once the presence of abnormal tissue is established, we may be able, through orderly investigations, to determine whether more specific manifestations of disease can be established.

NMR puts radiology on the threshold of one of its most exciting eras. The opportunities are tremendous because the directions in which it can be developed are numerous and uniquely inviting.

I

Principles

2

Basic Physics of NMR

William Bradley, M.D.

Heather Tosteson, M.F.A.

The phenomenon of NMR was first discovered in 1946 (1,2) and has since been used extensively as an analytic tool in chemistry and physics. Briefly, NMR is a process undergone by nuclei with magnetic moments when placed in a strong magnetic field. If these magnetized nuclei are exposed to radio waves of a specific frequency, they emit a signal of the same frequency as the one applied.

NMR was first used for internal imaging in 1973 (3). Since then, the medical applications of NMR imaging have rapidly been explored. The current excitement about the development of the technique reflects the facts that it is a noninvasive and nonionizing modality. In addition, intrinsic parameters of the technique yield information on the physiologic state and vascularity of tissues. Thus, NMR imaging holds great promise in the diagnosis of tumors, abscesses, and regions of edema and inflammation.

The basic physical principles of NMR that apply to medical imaging will be discussed in this chapter. We shall introduce the concepts of resonance, nuclear magnetic moment, nuclear angular momentum, magnetic relaxation times, and Fourier transforms and will attempt to provide conceptual models for these processes.

PHYSICAL CONCEPT OF RESONANCE

Let us first discuss the exchange of energy in a purely mechanical example.

Suppose we had a system composed of a box, a spring, a needle, an axle, a pulley, a string, and a weight and it was designed as follows: The needle is held upright by the spring, which is attached both to the tip of the

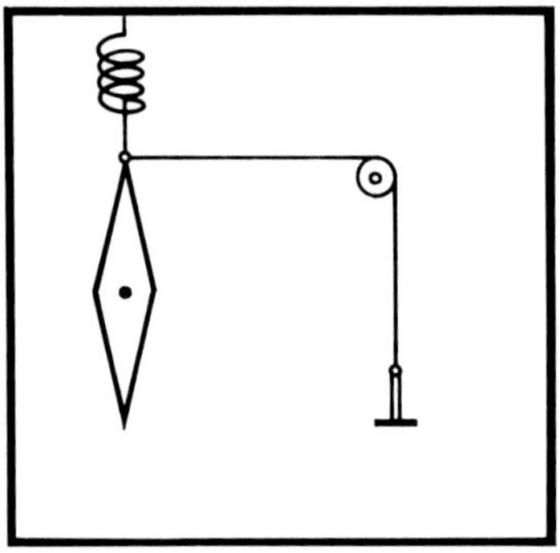

Figure 2.1. A needle held upright by a spring.

needle and to the ceiling of a wooden box. An axle is attached to the center of the needle. The string is attached to the needle and goes over a pulley to a weight (Fig. 2.1). The system as it stands now is perfectly balanced and in a state of equilibrium. Should we, however, apply force to the weight, the string would turn the pulley, which, in turn, would force the needle from the upright position (Fig. 2.2).

If the applied force is constant, for as long as it is applied to the weight, the deflected needle will remain in a second equilibrium position, which is determined by the additional energy added to the system. This can be regarded as a steady state situation: as long as the additional force is present, the second equilibrium state will be maintained. However, should the force be removed, the needle will return to the original upright position, doing work in the process, that is, drawing the weight back up to its original position. Ignoring friction, the amount of work done by the needle is equal to the force originally applied to the weight multiplied by the distance the needle was deflected from the upright position.

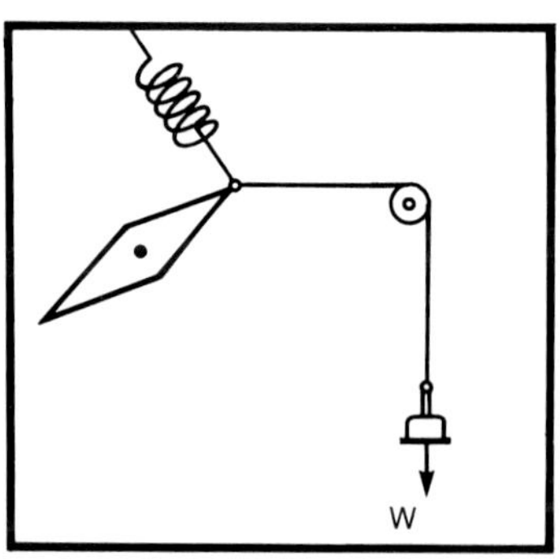

Figure 2.2. A weight (w) exerts a force that moves the needle to a new equilibrium position.

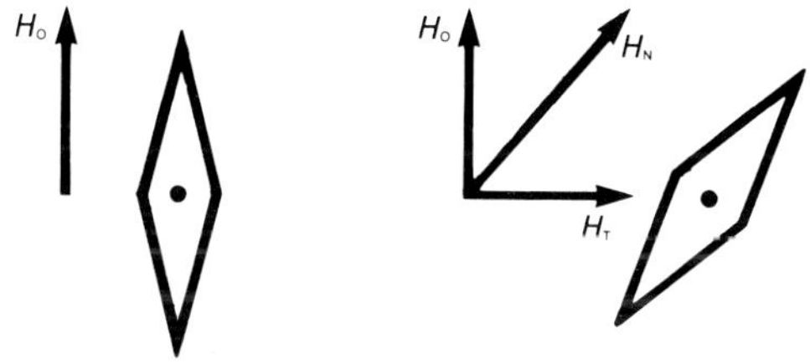

Figure 2.3. A magnetic needle can be held pointing up by a magnetic field H_o. Application of a second field (H_t) moves the needle to a new equilibrium position given by the direction of the net field (H_n).

Now, let us vary our example by introducing magnetic fields both for the force of the spring and for the additional force applied to the weight (Fig. 2.3). If the iron needle is placed in a strong magnetic field, it will become magnetized and align with the field. If we wish to deflect the needle from the upright position, we must apply a force. This force can be either mechanical or magnetic, so if we introduce a second magnetic field at an angle to the first field, we will be able to deflect the needle. The force will be determined by the strength of the magnetic field, the size of the needle (i.e., its magnetic dipole moment), and the desired angle of deflection.

The most efficient way to achieve maximum deflection of the needle is to apply the second field perpendicular to the first one. As in the original example, the needle, responding to the constant force of the second field, will move to a second equilibrium state, which will be maintained as long as the second force is applied. However, as soon as the second field is removed, the needle will return to its original equilibrium state, doing work in the process.

Let us refine the example further. Rather than assuming that an arbitrary amount of needle deflection is possible, let us add an additional constraint found in quantum mechanics. Let us imagine that there are only two positions that the needle can hold. Then, the work done to deflect the needle must be exactly equal to the energy difference between these two states, or there can be no exchange of energy between systems. Resonance is the term used to describe the process of the exchange of energy at discrete levels.

Atomic nuclei can, like the needle in the preceding example, have magnetic dipole moments. In the next section, we shall discuss the specific nuclear properties that determine the kind and strength of magnetic force that can create resonance in magnetically susceptible nuclei.

NUCLEAR PROPERTIES

Magnetic Moment

A moving charge will generate an accompanying magnetic field at right angles to the direction of movement of the charge (Figs. 2.4a and 2.4b). Because electrons are charged particles with an intrinsic angular momentum, they have a magnetic moment associated with the presence of a rotating charge. Like electrons, other atomic particles may possess an intrinsic angular momentum, or spin.

The proton, for instance, has a spin equal to that of the electron. Because it also is a charged particle, it, too, has a magnetic moment associated with this spin. Even though the neutron is electrically neutral, its component electrical charges are not uniformly distributed within its volume. Charges closer to the core rotate at a slower rate than do opposite charges further out, resulting in two magnetic fields at different strengths oriented in opposite directions. The magnetic fields do not completely cancel out each other, so there is a net magnetic moment for the neutron, although it is about half the strength of the magnetic moment of the proton.

Nuclei other than hydrogen (which is a single proton) are a combination, or "ensemble," of "nucleons" (the generic term for neutrons and protons). Such an ensemble can have net spin (i.e., magnetic moment) if it has either an odd number of protons or neutrons, or both. When pairs of protons inside a nucleus align with their spins opposite to each other, the

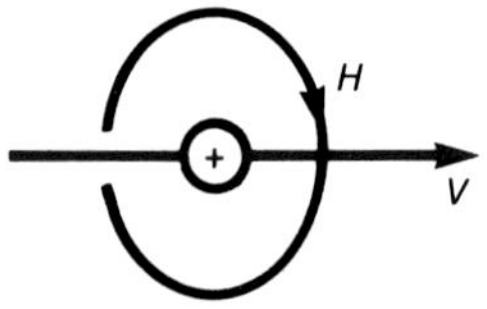

Figure 2.4a. A charge moving with a velocity v produces a circular magnetic field (H).

Figure 2.4b. A charged sphere rotating with an angular velocity ω produces magnetic field lines (H), which can be represented by a net vector (**m**) aligned with the axis of spin.

net spin for the pair is zero. Similarly, pairs of neutrons align so as to cancel their spins. Thus, it is only the odd proton or neutron, or both, that contributes a magnetic moment to the nucleus.

NMR

Nuclear spin quantum number values are either zero, an integer, or half integer. For example, the nuclear spin number of hydrogen is $\frac{1}{2}$, as it is for ^{31}P, whereas the nuclear spin number of ^{6}Li is $\frac{3}{2}$. From the spin number can be determined the number of energy states possible for the nucleus when placed in a steady magnetic field. This number is $2I+1$, where I is the nuclear spin number. Therefore, the number of possible energy states for hydrogen is two. Lithium has four possible energy states. For all elements, the separation between the highest and lowest energy levels in a steady magnetic field (**H**) is $2\mu\mathbf{H}$, where μ is the nuclear magnetic moment. The maximal separation is equivalent to the energy difference between the north and south alignments of the nucleus. The energy separation between different levels is thus $\mu\mathbf{H}/I$, and the frequency of electromagnetic radiation that can effect transitions between the states is $\mu\mathbf{H}/Ih$, where h is Planck's constant. Thus, a nucleus with two possible energy states is capable of making transitions between these states if it is exposed to a steady source providing electromagnetic energy equal to $2\mu\mathbf{H}$. The nucleus absorbs or emits energy as it changes its alignment, that is, as it resonates.

Because nuclei have angular momentum, they behave in a different way in a magnetic field than does the needle in our original example. Nuclei aligned in a magnetic field are not static. They behave, instead, like toy tops aligned with the earth's gravitational field: they wobble, or "precess," around the axis of the main field (Fig. 2.5). The frequency at which they precess depends on the strength of the magnetic field **H**.

The Larmor theorem provides an easy way to determine the precessional frequency ν of any particular atomic nuclei in a magnetic field of a given strength:

$$\nu = \left(\frac{\gamma}{2\pi}\right)\mathbf{H}, \qquad [1]$$

where γ is a constant called the magnetogyric ratio, which has a specific value for each atomic nucleus. (Note that Planck's constant has dropped out of the equation.) Because the magnetogyric ratio (and therefore the quantity in parentheses) is constant, the precessional frequency ν changes directly with the strength of the magnetic field **H**. The higher the field, the higher the frequency of precession.

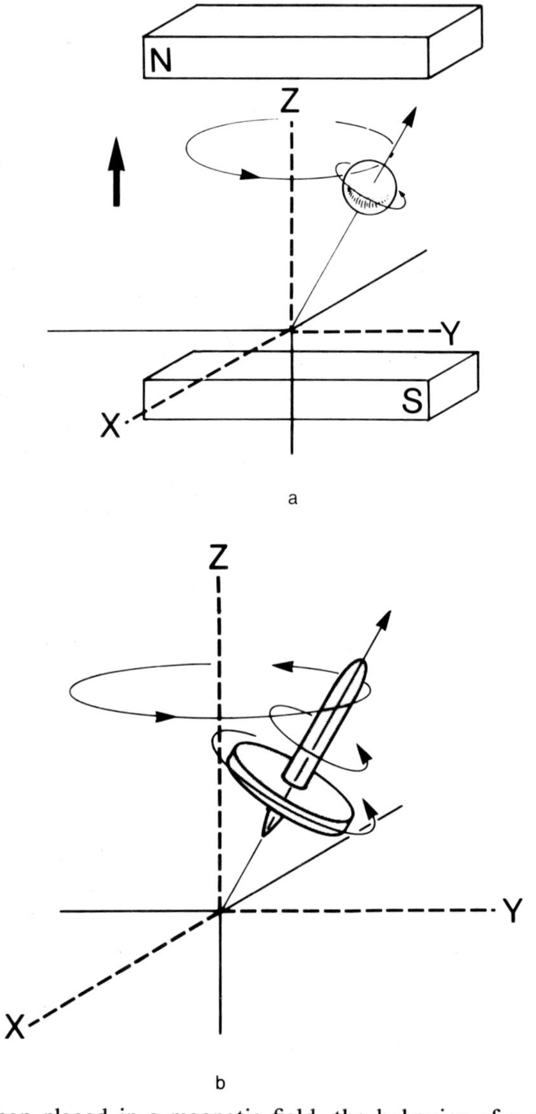

Figure 2.5. When placed in a magnetic field, the behavior of a nucleus with a magnetic moment (a) is similar to that of a spinning top on a table (b).

To apply a steady force to a precessing object, the force must oscillate at the same frequency at which the object precesses. Although, in principle, a rotating magnetic field could be created to provide this force, electromagnetic waves, which are composed of oscillating electrical and magnetic fields at right angles to each other, are used. The frequency ν of the electromagnetic radiation needed to induce resonance is identical to the precessing frequency of the nuclei, as determined by the Larmor theorem.

Electromagnetic radiation in the radio frequency range provides the spinning nuclei with the quantum of energy needed to effect transitions between energy states. The nucleus, moving from a low-energy state to a high-energy state, will absorb a photon; when moving from a high-energy state to a low-energy state, it will emit a photon. When the second magnetic field, namely, the radio wave, is present, the original state of equilibrium is replaced by a second steady state that reflects the new distribution of magnetic forces and the presence of additional energy in the system. When the disturbing force is removed from the system, the nucleus will emit energy as it returns to its state of original equilibrium. When hydrogen is placed in a magnetic field of 3.5 kG and exposed to a radio frequency field of 15 MHz, it will emit energy at the same 15-MHz frequency. These radio waves have wavelengths of 20 m.

CHEMICAL ENVIRONMENT

Net Magnetization: Equilibrium

So far, we have used a simple mechanical model to discuss the exchange of energy that constitutes nuclear magnetic resonance. We shall now add other factors to the discussion, paying particular attention to the complex molecular environment binding and surrounding the resonant nuclei. Although the intensity of the NMR signal depends on the density of the resonant nuclei, the duration of the NMR signal is completely determined by the chemical environment of the nuclei.

Even in a magnetic field, atoms are constantly in thermal motion, interacting with other atoms. To borrow an analogy from House (4), the protons in a magnetic field are like compasses lying on a table, all pointing toward the north pole. Adding the effect of thermal motion is like shaking the table. The compasses swing around and point in every conceivable direction. Yet when all are averaged together, the net direction of the compass needles will still be toward the north pole; that is, the average vector of the agitated compass needles always points north.

The same concept holds true for a beaker of water in a magnetic field. Each water molecule contains two hydrogen nuclei that tend to line up with the magnetic field. Thermal motion will cause the magnetic moments of most of the protons to point randomly, but the average will still be toward the north pole. The average of all these magnetic moments is called the magnetization **M.** Possessing magnitude and direction, this parameter can be described as a vector. Once equilibrium has been achieved, the magnetization **M** points straight toward the north pole of the magnet with its maximal (equilibrium) value $\mathbf{M}_o$.

The net magnetization is small, approximately 1.4 ppm, that is, 1.4 excess protons pointing north per million protons. The equilibrium ratio can be expressed as follows:

$$N\,(\mathrm{Up})/N(\mathrm{Down}) = \exp(-\mu \mathbf{H}_o/kT)$$
$$= 1 + 1.4\times10^{-6} \text{ for } 37°\mathrm{C} \text{ and } 3.5 \text{ kG}, \quad \mathbf{(2)}$$

where k is Boltzmann's constant, and T is temperature in degrees Kelvin. It is this small excess of protons in the lower-energy alignment that we are exclusively concerned with because these protons are the only ones capable of being manipulated by the intrusion of energy into the system. It is important to remember, however, that the net magnetization, the vector **M,** is a mathematical concept and that it reflects the sum of a variety of different nuclear motions (Fig. 2.6). The net response alone is observable and is what will be described in the following sections. How-

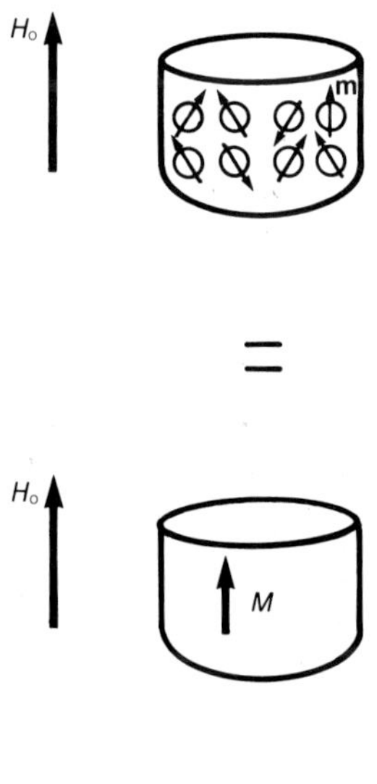

Figure 2.6. The net magnetization (M) of a sample results from summing the individual magnetization vectors (**m**).

ever, it is useful to keep in mind the variety of individual nuclear motions that lead to the observed bulk effects.

Transverse Magnetization: Resonance

If the magnetization is disturbed from equilibrium, it behaves like an individual hydrogen nucleus, precessing around the axis of the external magnetic field. The frequency of the precession is predicted by Larmor's theorem and, again, depends on the strength of the magnetic field **H**. While the second magnetic field is being applied as a radio frequency pulse, the magnetization **M** precesses further and further away from the z axis, depending on the energy and duration of the pulse (Fig. 2.7). After stimulation, the nuclei, returning to equilibrium, lose energy by emitting radio waves and by transferring energy to surrounding molecules. The process by which energy is lost to the environment is called "relaxation."

For clarity, we can think of the magnetization vector **M** as having a component $\mathbf{M}_z$ along the z axis (which we define as the direction of the field $\mathbf{H}_0$) and a component $\mathbf{M}_{xy}$ perpendicular to the field (Fig. 2.8). $\mathbf{M}_z$ is called the "longitudinal magnetization," and $\mathbf{M}_{xy}$ the "transverse magnetization." Thinking of **M** in this way allows us to discuss the separate effects of the two external magnetic forces. At equilibrium, $\mathbf{M}_z$ is equal to $\mathbf{M}_0$ and is equivalent to the net magnetization, whereas $\mathbf{M}_{xy}$ is zero. When the sample is being excited by a radio frequency pulse, $\mathbf{M}_z$ decreases as

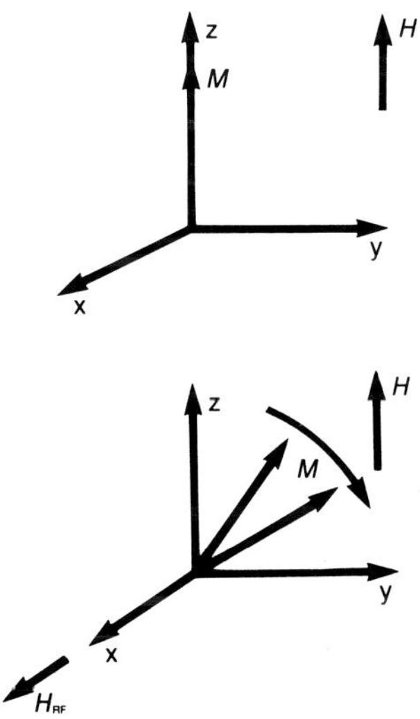

Figure 2.7. The magnetization vector **M** is initially aligned with the field H. When a radio frequency field is turned on, the vector **M** precesses away from the z axis.

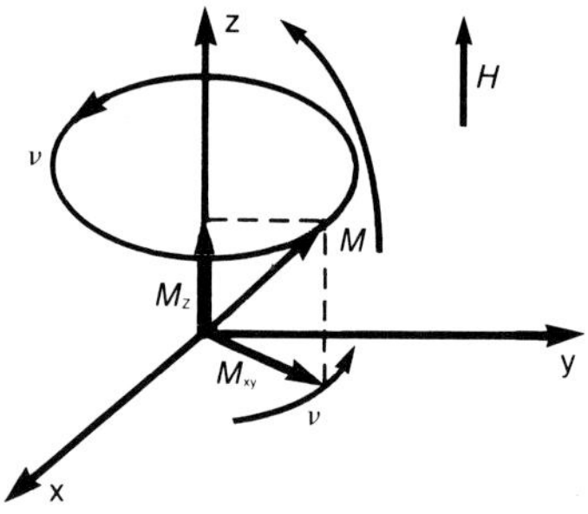

Figure 2.8. After the radio frequency field is turned off, **M** returns to the equilibrium position. The transverse magnetization M_{xy} decays to zero in the process. The vector $\mathbf{M}_{xy}$ precesses around the z axis with a frequency ν.

$\mathbf{M}_{xy}$ increases and precesses around the z axis at a frequency ν. After the radio frequency pulse, $\mathbf{M}_z$ returns to $\mathbf{M}_0$, where $\mathbf{M}_{xy}$ decays to zero. These processes are exponential in time, much like the decay of radioactive material. They are characterized by two time constants, T_1, the longitudinal magnetic relaxation time, and T_2, the transverse magnetic relaxation time.

Return to Equilibrium: T_1 and T_2

As **M** returns to equilibrium, its length is not constant; that is, the strength of the precessing magnetization vector is not constant. In general, the transverse magnetization is lost more rapidly than longitudinal magnetization is regained; that is, T_2 is generally shorter than T_1. T_1 and T_2 depend on the molecular structure of a sample, its physical state (i.e., liquid or solid), and on the strengths of the internal magnetic fields. Whether a sample is solid or liquid depends on whether the forces between molecules are fixed or fluctuating. These forces affect the rate of exchange of thermal energy by determining the frequency of collision between molecules. The presence and relative strengths of the internal magnetic fields affect the rates of response of the resonating nuclei to the external magnetic fields.

The longitudinal magnetic relaxation time, T_1, measures the time it takes the sample initially to become polarized in the fixed external magnetic field $\mathbf{H}_0$. It is also the time constant of return to equilibrium in the fixed magnetic field after a radio frequency pulse. The rate of return to equilibrium along the z axis depends on how fast energy can be transferred from the spinning protons to the rest of the lattice; therefore, T_1 is sometimes called the "spin lattice relaxation time." Because energy is transferred to the environment by thermal interactions, that is, by random thermal collisions between molecules, T_1 is also sometimes called the

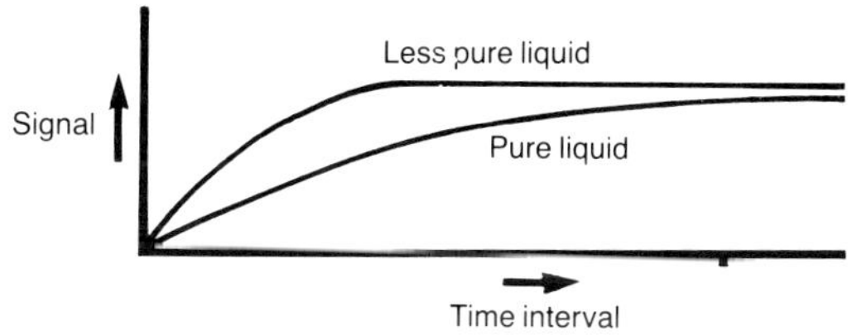

Figure 2.9. Signal obtained as a function of the interval between contiguous excitations of the sample is exponentially dependent on the T_1 value. T_1 for solids is long compared to that for liquids. Very pure liquids have longer T_1 values than those obtained when the liquid has impurities (such as proteins) added to it.

"thermal relaxation time." T_1 is shorter in a liquid (seconds) than in a solid (minutes to hours) (Fig. 2.9). Magnetic redistribution occurs more quickly in liquids because the fluctuation of molecules allows for the rapid transfer of thermal energy.

T_2 is a measure of how long the resonant nuclei hold the temporary transverse magnetization $\mathbf{M}_{xy}$. It indicates the relationship between the strength of the external field and the strength of local internal fields. For precessing protons, "holding the transverse magnetization" means wobbling or oscillating in phase. After a radio frequency pulse, the protons are synchronized and begin precessing together. However, if they experience locally different magnetic fields, they will precess at different frequencies, quickly getting out of phase, losing "coherence." When protons precess at slightly different frequencies, the collective transverse magnetization soon averages to zero.

As mentioned previously, the precessional frequency of a group of protons is determined by the strength of the magnetic field at a particular point. At any moment, the magnetic field is the sum of the external magnetic field and any local fields that may be present. In a liquid, the local magnetic fields from neighboring molecules fluctuate rapidly. They thus average to zero and contribute little to the net magnetic field at any point. Solids, however, have fixed molecules with fixed magnetic fields. These internal magnetic fields are significant and cause local variations in the value of the magnetic field, resulting in a rapid loss of coherence. Thus, for solids, T_2 is very short (microseconds). For liquids, the internal fields are weaker, and coherent (in-phase) transverse magnetization is maintained for a longer time (seconds) (Fig. 2.10). Because the exchange of energy from one spinning proton to another causes dephasing, the transverse magnetic relaxation time T_2 is also called the "spin-spin relaxation time."

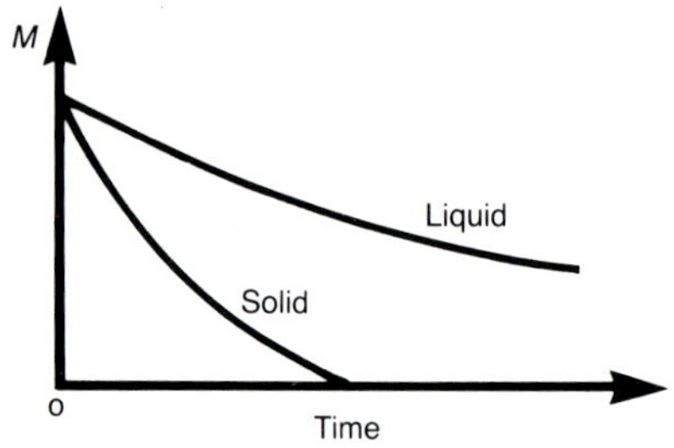

Figure 2.10. The magnetization M decays as a function of time in an exponential manner ($M = M_0 \exp - (t/T_2)$). The T_2 values of liquids are longer than those of solids.

NMR SIGNAL

The degree of induced transverse magnetization is controlled by the strength of the radio frequency radiation and will be reflected in the amplitude of the signal evoked from the resonant nuclei. (The frequency of the radiation will be constant for a given element in a given magnetic field.) The signals most commonly evoked in NMR imaging are the continuous wave, the free induction decay, and the spin echo.

Continuous Wave

If a low-amplitude radio frequency wave is continuously imposed on the main magnetic field, the nuclei will absorb energy from the radio frequency field and emit energy in the form of radio frequency radiations and thermal motion at a rate that maintains the net magnetization at a constant angle from the z axis. Although the transverse magnetization induced by the radio frequency field has a persistent tendency to decay while in the main field, enough energy is continuously introduced into the system to maintain a constant degree of transverse magnetization.

The resonant signal is perceptible only when the radio frequency field is turned off. Therefore, the continuous-wave signal is not measured as a flow of radio frequency radiation from the nuclei; instead, it is observed as a steady loss of radio frequency energy from the radio frequency source.

Free Induction Decay

If the nuclei are exposed to a somewhat stronger, medium-amplitude radio frequency signal, energy is added to the system faster than it can be radiated away. The net magnetization precesses at an increasingly greater angle to the z axis. If enough energy is added to bend the magnetization 90 degrees away from the vertical z axis, **M** will precess in the horizontal xy plane. This burst of radio frequency energy is called a 90-

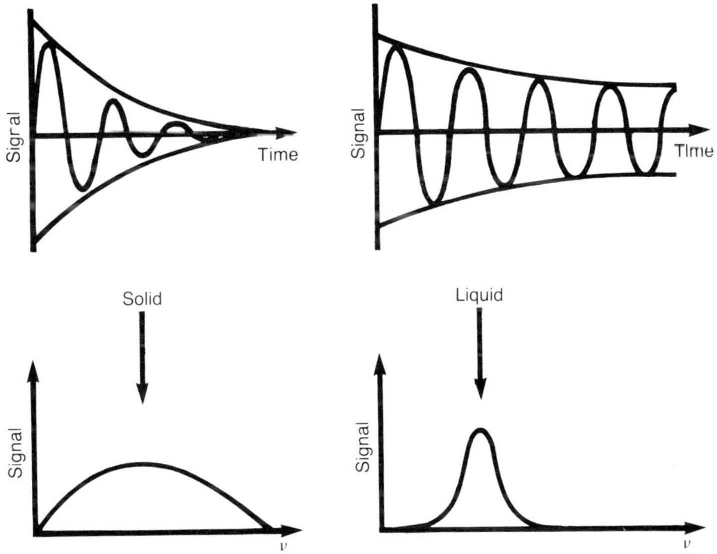

Figure 2.11. The time constant of the envelope of the signal is given by T_2. Samples with a short T_2 contain a broader range of frequencies than those in a sample with long T_2. The time-domain signal is converted to the frequency domain by a Fourier transformation.

degree pulse. Once the radio frequency pulse is over, the nuclei, precessing back to their original alignment, will emit energy in the form of radio frequency radiation. The signal is observable as long as the nuclei precess in phase. The decay curve of the signal is called the free induction decay (FID). If $\mathbf{H}_0$ is perfectly uniform, the time constant of the decay curve is T_2 (Fig. 2.11).

Solids and liquids exhibit different FIDs when placed in a homogeneous magnetic field. The FID for solids is composed of sine waves of many different frequencies, resulting in a short signal of erratic amplitude. The erratic signal is due to the presence of significant local magnetic field inhomogeneities that cause nuclei in close proximity to precess at slightly different frequencies. The longer FID for liquids can be represented as a single sine wave, or as superimposed sine waves of almost the same frequency, because of the greater coherence of the precessional frequencies of the individual nuclei (Fig. 2.11).

The FID, observed as a single current oscillating in the radio frequency range, can be represented by its component frequencies by use of the Fourier transformation. The observed NMR signals from a solid and a liquid and the different frequencies that compose each of these oscillating signals are shown in Figure 2.11. Notice that, as the different frequencies become increasingly out of phase, the amplitude of the overall signal decreases.

Spin Echo

Another NMR response, the spin echo, is created by sequentially imposing two pulses of different amplitudes: first a 90-degree radio frequency pulse, then a 180-degree pulse. Exposed to the 90-degree pulse, the nuclei will behave as described in the discussion of an FID. They begin by precessing in phase and lose coherence quickly due to local field inhomogeneities caused by nonuniformity of the external field $\mathbf{H}_0$ and the presence of significant internal fields in the substance. If the direction of the net transverse magnetization is now reversed by exposure to a 180-degree pulse, a different NMR signal called the "spin echo" will result (Fig. 2.12). Although the frequencies of the resonating protons remain the same, their relative phase shifts will be reversed, resulting in a signal that is a bell curve. The first half of the curve of the spin echo signal is a mirror image of the original FID signal as the precessing nuclei regain a state of coherence. The second half of the spin echo signal duplicates the original FID curve after the 90-degree pulse, although with slightly less intensity than the original signal. This difference is due to the irreversible loss of coherence due to the internal field variations. In Figure 2.13, the dashed curve indicates T_2 decay due to loss of coherence from internal field variation. The solid curves include the loss of coherence due to nonuniformity of the external field $\mathbf{H}_0$, and represent the observed signal.

CREATING THE NMR IMAGE

Gradients

So far, we have discussed how to evoke an NMR signal from a substance in a relatively homogeneous magnetic field. This is similar to NMR spectroscopy, which uses high and extremely uniform magnetic fields to analyze the chemical structure of small quantities of a homogeneous substance. Because there is no spatial variation in the field, a homogenous substance should yield an identical signal from the entire sample. Variations in the NMR signal, which creates the frequency spectrum for a particular substance, will thus give evidence of fine molecular structure, as discussed in Chapter 8. Figure 2.14 gives the NMR spectrum of hydrogen for the ethanol molecule.

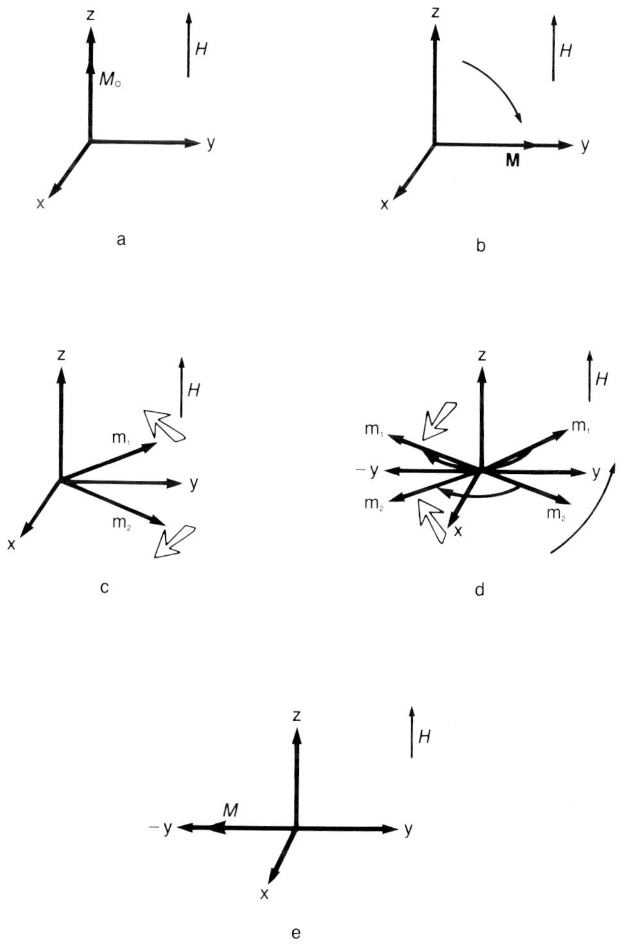

Figure 2.12. Production of a pulse echo. (a) M_0 gives the initial equilibrium condition of the sample. (b) A radio frequency pulse rotates the magnetization vector (**M**) onto the y axis. (c) Individual nuclei (m_1, m_2) precess around the z axis at different rates and lose phase coherence. (d) A 180-degree radiofrequency pulse reverses the y component of individual vectors, so that they now start to come into coherence. (e) Coherence is achieved when the individual vectors align, producing a maximum magnetization M.

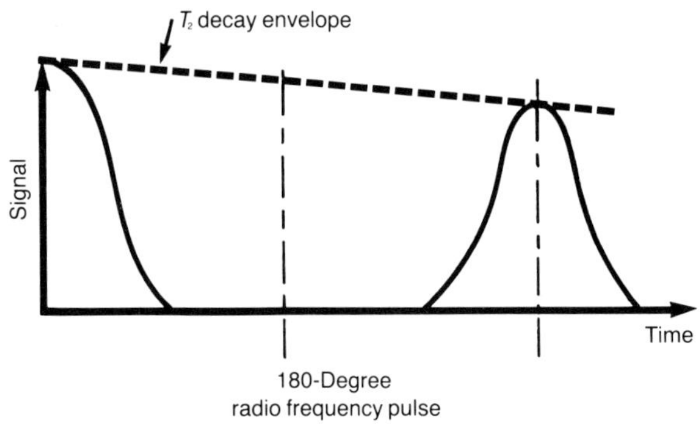

Figure 2.13. Pulse echo occurs after a 180-degree pulse. Peak value decays with a T_2 exponential time (dashed curve).

NMR imaging techniques differ from NMR spectroscopy in the use of magnetic field gradients to provide spatial resolution for the NMR signal. Because the value of the magnetic field determines the resonant frequency, if a magnetic field is created with a different value at every point, excited nuclei at different positions in the sample will resonate at different radio frequencies. Resonant frequencies will then encode spatial information, and the frequency spectrum will describe a spatial distribution.

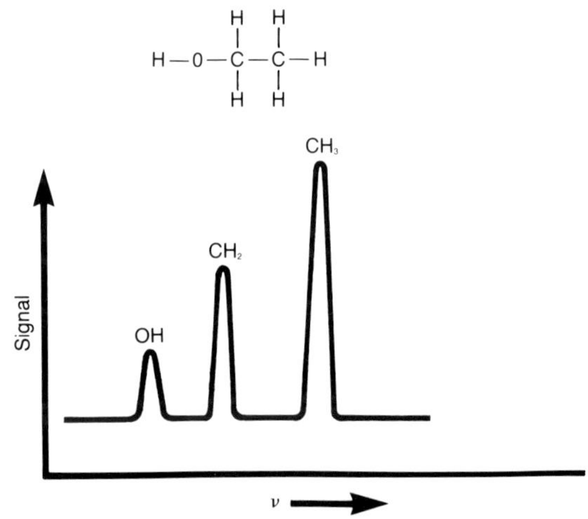

Figure 2.14. Simplified NMR spectrum of the hydrogen in ethanol. One hydrogen atom is bound to oxygen (OH), two are bound to the median carbon, which binds to both the oxygen and the other carbon atom (CH_2), and three are bound to the second carbon, which binds only to hydrogen and to the preceding carbon (CH_3).

Because it is impossible to create a static magnetic field with a different strength at every point in space, variable magnetic fields are used to define an imaging volume. In almost all techniques, three different magnetic field gradients are employed either during or after the application of the radio frequency radiation. Each gradient (i.e., variation in magnetic field along a particular axis) will define a plane of the sample in which the magnetic field is the right strength to allow resonance at a specific radio frequency. By applying additional gradients, the resonant volume can be further reduced from a plane to a line or a point.

Once the magnetic field is defined at a particular point in time, analysis of the composite frequencies of the NMR signal provides information about spatial distribution of resonant nuclei. The amplitude at each frequency provides information about the density of resonating nuclei. These processes are discussed in detail in Chapter 2.

Fourier Transformation in NMR Imaging

Consider three identical water samples in a perfectly homogeneous magnetic field with a continuous linear gradient superimposed on it (Fig. 2.15). After excitation with a broad-band radio frequency pulse (a pulse containing several different radio frequencies), protons in each region of the magnetic field will emit radio waves of a fixed frequency, decaying in intensity in an exponential manner. An antenna will sum these three signals into a more complex wave. This signal (called the time-domain signal) encodes the frequencies that compose it in a manner not directly amenable to observation. The Fourier transformation is a process whereby the signal is transformed from the time domain to the frequency domain. Stated differently, it is a representation of the signal by the relative amplitude of its component frequencies during the period of observation.

For the case given here, the transformation results in three distinct frequencies (Fig. 2.15). If the water were spatially distributed uniformly throughout the field, all frequency bands would contain a signal. Knowledge of frequency is equivalent to knowledge of magnetic field strength from Equation 1. Since the spatial variation in field across the volume is known, the three frequency values obtained from the Fourier transformation indicate the positions of the three water samples (Fig. 2.15).

NOTE

It should be noted that both quantum and classical physical concepts are commonly used in descriptions of NMR. Quantum mechanical concepts,

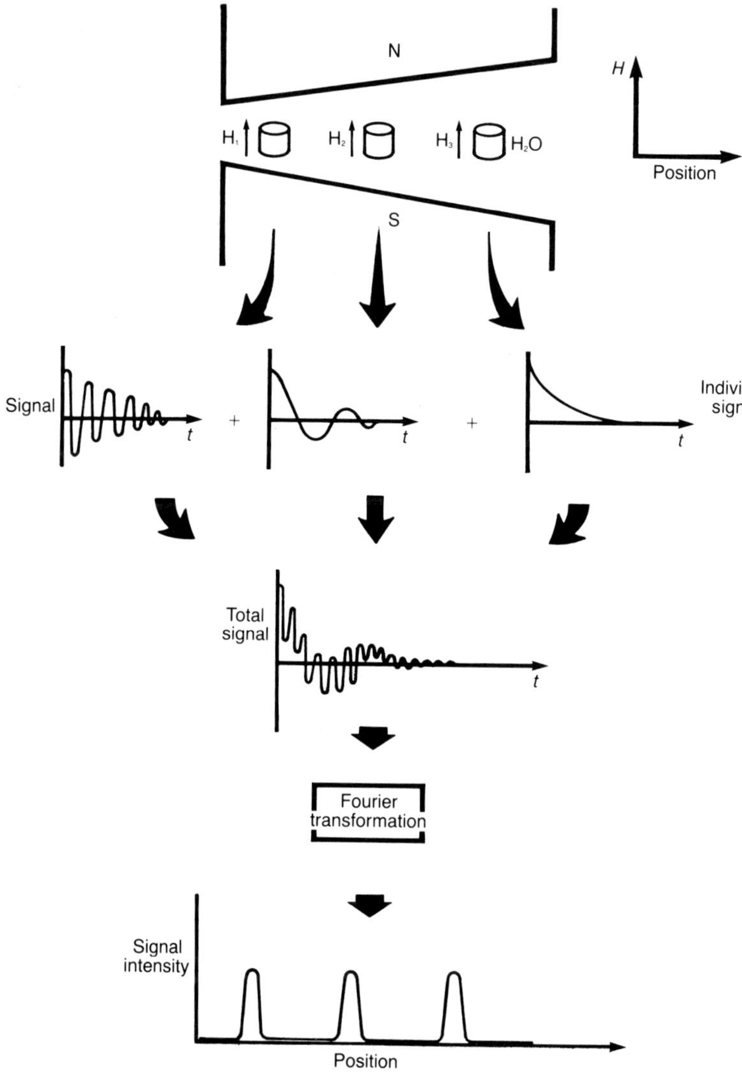

Figure 2.15. Steps for obtaining spatial information in NMR are shown from top to bottom. Samples are placed in a magnetic field that varies in strength along the horizontal axis. The signals obtained from each position in the magnetic field have a frequency that is directly proportional to the strength of the field. The summed signal that is observed contains these different frequencies, which are recovered by a Fourier transformation (FT). The output of the FT is a plot of signal intensity versus frequency. Because the value of the magnetic field along the axis is known, each frequency value can be assigned to a unique position.

like spin numbers and discrete energy states, are indispensable when discussing resonance, as are the equations that determine the exact frequencies that will induce resonance. When a visual model is needed, however, classical physical concepts—vectors and precession—are convenient to use. Both models yield equivalent results because Planck's constant drops out from the resonance equation (refer to Equation 1). Therefore, by definition, because the quantum of action is no longer present, the quantum mechanical results should also be obtainable from a classical model (5). The concurrent use of the two models results from the attempt to express the full conceptual complexity of the NMR process in a manner comprehensible to the nonspecialist. For readers who are interested in a more thorough discussion of NMR, we have included a short bibliography of standard books on the subject.

REFERENCES

1. Bloch F: Nuclear induction. *Phys Rev* 70:460–473, 1946.
2. Purcell EM, Torrey HC, Pound RV: *Phys Rev* 69:127, 1946.
3. Lauterbur PC: Image formation by induced local interactions: Examples employing nuclear magnetic resonance. *Nature (London)* 242:190–191, 1973.
4. House WV: Introduction to the principles of NMR. *IEEE Trans Nucl Sci* NS-27:1220–1226, 1980.
5. Slichter CP: *Principles of Magnetic Resonance*. New York, Harper & Row, 1963, pp 1–9.

BIBLIOGRAPHY

Abragam A: *The Principles of Magnetic Resonance*. New York, Oxford University Press, 1961.

Andrew ER: *Nuclear Magnetic Resonance*. Cambridge, Cambridge University Press, 1969.

Bloembergen M: *Nuclear Magnetic Relaxation*. New York, WA Benjamin, 1970.

Carrington A, McLachlan AD: *Introduction to Magnetic Resonance*. New York, Harper & Row, 1967.

Farrar TC, Becker ED: *Pulse and Fourier Transform NMR*. New York, Academic Press, 1971.

Schumacher RT: *Introduction to Magnetic Resonance*. New York, WA Benjamin, 1970.

Slichter CP: *Principles of Magnetic Resonance*. New York, Harper & Row, 1963.

3

Overview of NMR Imaging Techniques

Lawrence E. Crooks, Ph.D.

Three basic technical considerations have governed the development and refinement of NMR imaging techniques: how to design magnets that produce a uniform field over a human size volume, how to determine an imaging volume, and how to analyze the NMR signal. Achieving spatial resolution has posed an important challenge because it has required the greatest technical departure from conventional NMR spectroscopy as well as from all optically based imaging methods. Solutions to this problem have, of necessity, been the focus of almost all NMR imaging research until now. The number of NMR imaging techniques that have already been developed—at least six, with variations on each (1–6)—illustrates the diversity of possible approaches to the problem of volume selection. Techniques are commonly grouped according to the kind of volume they choose to select, namely, single-point, line, plane, or whole-object techniques. In this chapter, I shall discuss these groupings of existing imaging techniques and will include both a description of the method by which the volume is selected and an indication of the effects of the imaging method on the kind and quality of data collected.

In the future, as images derived from any one imaging system improve in spatial resolution, signal analysis will become increasingly important. Even now, any comparative description of images derived by different techniques illustrates the complex interdependence between the technique used to induce resonance and the resultant signal. The numerous and complexly interwoven physical and chemical characteristics reflected in the NMR signal increase NMR's unique potential for producing images with a high level of chemically significant soft-tissue contrast.

Signal intensity depends on the number of resonating nuclei in the volume element, their motion through the volume, their magnetic relaxation time constants T_1 and T_2, their chemical state, and electronic noise. With the exception of electronic noise, all the components of intensity seem useful for identifying structure or function in the body. The potential of NMR imaging is the amount of information available from these components of the signal intensity. The challenge of NMR imaging is to extract these components in a coherent manner and to correlate them with body function.

The extent to which an NMR imaging technique can mask or magnify the factors that contribute to signal intensity increases the challenge of, and necessity for, extensive and systematic research in tissue characterization with any one imaging system and simultaneous research to determine the interrelationship between each imaging technique and its resultant signal. Chapter 5 contains a discussion of initial research on tissue characterization with a single imaging system. In the discussion of imaging techniques, I try to indicate how the relative contributions of the T_1 and T_2 parameters to the NMR signal will differ with the different imaging techniques, thus resulting in appreciable variations between the intensity images derived by different techniques.

Certainly, all NMR imaging techniques need to be considered in terms of the basic performance characteristics by which the efficiency and the applicability of any diagnostic method are eventually evaluated. Questions that need to be posed include the following: In general, how do the speed and resolution of NMR imaging compare with the speed and resolution of x-ray computed tomography, nuclear medicine, or ultrasound? Which NMR imaging techniques require longer imaging times? How sensitive is any particular imaging technique to subject motion? What are the fundamental limitations to NMR imaging refinement due to intrinsic signal-to-noise considerations? How easy will NMR systems be to operate? A brief discussion of these issues will conclude the chapter.

IMAGING PROCESS

NMR Techniques

To explain how signal intensity depends on relaxation times, we need to describe in greater detail the basic ways signals can be observed. As in the previous chapter, the discussion will focus on hydrogen because it is the element normally used for NMR imaging.

FID

As described in Chapter 2, the nuclei in an object will align along the direction of an applied magnetic field. When an object is in a magnetic field for a sufficiently long time, an equilibrium is reached. A net nuclear magnetic field, which is the sum of the magnetic moments of all the nuclei, is induced in the object. By applying a radio frequency magnetic field at the resonant frequency, this net nuclear magnetic field can be tipped away from alignment with the main field. The amount of tipping depends on the strength and duration of the radio frequency magnetic field. The portion of the nuclear magnetic field that is in the plane of the axis of the radio frequency coil will induce a radio frequency voltage in the coil. This voltage can be observed after the field is turned off. The observed voltage is called a free induction decay (FID) because it is free of the tipping radio frequency field; it is a voltage induced in the coil by the net transverse nuclear magnetic field, and it is a decaying voltage because the net nuclear transverse magnetic field decays as the nuclei return to their equilibrium orientation with the main field. The transverse nuclear magnetic field decays to zero with a T_2 time constant; thus, the FID decays to zero with the T_2 time constant.

When all the net nuclear magnetic field is tilted into the plane of the radio frequency coil, the signal can be described as: Signal α $\mathbf{M}_0$ $\exp(-t/T_2)$, where t is the time elapsed from the start of the FID, and $\mathbf{M}_0$ is the net magnetic moment at equilibrium. For this amount of tip, the return to equilibrium of the nuclear magnetic field along the main field can be described as

$$\mathbf{M}(t) = \mathbf{M}_0 \, [1 - \exp(-t/T_1)]. \qquad [1]$$

If the nuclear magnetic field had been tipped over so that it is directed opposite the main field (i.e., inverted), no signal will be observed because there is no magnetic field in the plane of the radio frequency coil, and the return to equilibrium of the nuclear magnetic field can be described as

$$\mathbf{M}(t) = \mathbf{M}_0 \, [1 - 2 \exp(-t/T_1)]. \qquad [2]$$

T_1

Note that there is no way to observe the T_1 return to equilibrium along the main magnetic field using a single FID because you receive no direct signal from magnetization aligned along the main field. By sampling the nuclear recovery process starting from some known nonequilibrium position, you can measure T_1 indirectly. Thus, NMR parameters are commonly measured by using a sequence of operations, in this case the application of two radio frequency pulses to tip the net nuclear magnetic field and the observation of at least one FID.

For example, a two-FID sequence used to determine T_1 could go as follows: First the nuclear magnetic field is tipped into the plane of the radio frequency coil, then the FID is recorded; after a time t_1 (seconds), the procedure is repeated. During the waiting period, t_1, the nuclei return toward equilibrium along the main field. If t_1 is less than about $4T_1$ of the sample, not all nuclei reach equilibrium. Therefore, the signal in the second FID is reduced by $[1 - \exp(-t_1/T_1)]$, which will be the ratio of the starting levels of the two recorded FIDs. If the FIDs are called FID_1 and FID_2, a computer can calculate T_1 using the equation

$$T_1 = t_1/\ln[FID_1/(FID_1 - FID_2)]. \quad [3]$$

Also, FIDs for several values of t_1 can be used in a curve-fitting program more accurately to determine T_1.

Other measurement sequences differ mainly in the amount of radio frequency tipping used; thus, they have slightly different equations to solve.

Spin Echo

The shape of the FID would lead one to expect that T_2 could be computed by curve fitting to the FID. This is true only if the main magnetic field is perfectly uniform. In a nonuniform field, nuclei at different positions resonate at slightly different frequencies. These different frequencies produce destructive interference, causing a more rapid decay of the FID than would be expected. The destructive interference can be reversed by applying a sequence of radio frequency tips that invert the nuclear magnetization. If, at a time t_2, one applies a radio frequency pulse that inverts the nuclear magnetic field, the inversion causes the interference to become constructive, so t_2 seconds after the inverting pulse, the induced voltage in the coil again increases to a maximum. The interference continues and is again destructive, that is, causes the induced voltage to decay like an FID.

This regeneration of the signal, with its peak at time $2t_2$, is called a spin echo. Note that irreversible T_2 decay occurs during this process. The relation of the peak height of the spin echo $[\mathbf{A}(2t_2)]$ to the peak height of the FID $[\mathbf{A}(o)]$ is

$$\mathbf{A}(2t_2) = \mathbf{A}(o)\exp(-2t_2/T_2). \quad [4]$$

T_2 can be calculated by recording the FID and spin echo and using the previous equation. Applying additional radio frequency inversions at times $t = 3t_2, 5t_2 \ldots$ causes additional spin echoes whose peaks are at times $t = 4t_2, 6t_2 \ldots$. The number of echoes observed is limited by the T_2 decay, but when several can be recorded, they can be used to calculate T_2 by fitting the signal amplitudes to the equation

$$T_2 = 2nt_2/\ln[\mathbf{A}(o)/\mathbf{A}(2nt_2)] \quad [5]$$

for $n = 1,2,3 \ldots$. If this sequence occurs every t_1, the signal strength of the first spin echo is

$$\text{Signal } \alpha \ \mathbf{M}_0 [1 - \exp(-t_1/T_1)] \exp(-2t_2/T_2), \qquad [6]$$

that is, a combination of the T_1 and T_2 dependencies.

Continuous Wave

Free induction decays and spin echoes are two forms of signal by which the nuclear magnetic field can be observed. A third form of nuclear magnetic field observation is steady-state, or continuous-wave, observation. In this case, the net nuclear magnetic field is periodically tipped by radio frequency magnetic field pulses. The time between pulses is short compared to T_1 and T_2, and the signal is observed between pulses. This sequence causes the FID from each pulse to merge with the spin echoes from previous pulses and form a continuous signal. The continuously observable signal has an advantage because it provides a steady, rather than transient, source of energy. However, because the nuclei are never allowed to return to equilibrium with the main field, the transverse nuclear magnetic field is reduced compared to the field that produces the FID or the spin echo signal, and the continuous-wave signal is thus reduced in strength. Also, the relatively simple exponential dependence of signal on T_1 and T_2 is replaced by a more complex dependence, so T_1 and T_2 can be deduced from this technique only by varying the amount of tip that the radio frequency pulses cause and then solving a complex equation.

Volume Selection

The above techniques are used to observe signals from an entire object, both in NMR imaging techniques and in high-resolution spectroscopy. In NMR imaging, in addition, one wants to distinguish signals from distinct volume elements in the object. The basic concept underlying all techniques to achieve spatial resolution is that the resonant frequency of the nuclei is proportional to magnetic field strength. This is explicitly stated by the Larmor relation

$$f = \gamma \mathbf{H}, \qquad [7]$$

where γ is the magnetogyric ratio of the nucleus and is 42.58 MHz/Tesla for hydrogen. So, by making the magnetic field in an object vary with position, one makes the resonant frequency of the nuclei vary with position. By knowing the position dependence of the magnetic field, the position of resonating nuclei is defined by their resonant frequency. I shall describe later the many varied ways of tailoring magnetic fields and NMR sequences to create the desired image but will provide a simple example here to illustrate the concept.

Consider an object in a magnetic field that is made to vary by using what is known as a linear field gradient. In this case, the field is

$$\mathbf{H} = \mathbf{H}_0 + Gx, \qquad [8]$$

where **H** is the field, $\mathbf{H}_0$ is the main field value, x is position along the x axis of the object, and G is the strength of the gradient that defines the field variation. The resonant frequency of the nuclei as a function of position along x is

$$f = \gamma\, \mathbf{H}_0 + \gamma\, Gx. \qquad [9]$$

f, $\mathbf{H}_0$, γ, and G are all known constants, so position as a function of resonant frequency is

$$x = [(f/\gamma) - \mathbf{H}_0]/G. \qquad [10]$$

Resonant frequency must be measured to determine position. Resonant frequency is measured by recording the NMR signal, either FID, spin echo, or continuous wave, as a function of time and then using the mathematical relation known as the Fourier transformation to find the signal's frequency spectrum. The Fourier transformation analyzes the time-dependent signal's shape and finds the strength of each of the signal's component frequencies. The result is the relation of signal strength to frequency. Equation 10 allows the NMR frequency versus signal strength spectrum to be translated to a relation of intensity to position.

NMR IMAGING TECHNIQUES

NMR imaging techniques can be classified into four types, depending on the type of volume that produces the received signal used to make an image: a single point, a line, a plane, or an entire three-dimensional object. All techniques produce images resolved to single points, but when these points are defined in the imaging sequence varies. The single-point method determines the points in the image most directly, by measuring points one at a time; on the other hand, three-dimensional reconstruction repeatedly measures all points simultaneously from many viewing directions and calculates the signal contributed by each point in the object. The line and plane techniques fall between these extremes.

Single-Point Techniques

In the single-point techniques, the main, uniform field is modified, so that resonance is observed from only a single point, a small (e.g., 27 mm^3) volume element. An image is made simply by moving the sensitive point with respect to the object. This is a very direct way to produce an image and does not require a computer for image generation, but it certainly benefits from computerized image processing.

Since resonance is localized to a point, all the standard techniques developed for chemical NMR on isolated samples can be directly applied. Measurements of T_1, T_2, and hydrogen concentration, as well as localized chemical analysis (described in Chapter 8), can be made at the selected point. Making chemical measurements at all points would create prohibitively long imaging times; therefore, this localized chemistry capability of the technique would usually be applied to selected points of interest after an imaging scan has been completed. In addition, if the point is positioned at a blood vessel, it is possible to acquire the signal-versus-time plot for pulsatile blood flow (7). Using continuous or gated observation, one could stop motion at a point, although present techniques, by averaging over the motion, usually blur movement to an average value. The single-point imaging volume limits motion artifacts to blurring because motion modifies only the intensity of the point through which something moves.

The main problem with single-point techniques is that they are slow. In the time that it takes to observe one point, all the points along a line could be observed with the same signal to noise. Speed would increase by the number of points in the line.

Two methods of defining sensitive points have been developed. As with most imaging techniques, their feasibility was first demonstrated by modifying standard instrumentation used in NMR spectroscopy. One method shapes the main magnetic field, so that the nuclei in a star-shaped region are at the resonant frequency and the nuclei outside this region are not at the resonant frequency. The center of the star is the sensitive point; the points of the star taper rapidly and contribute a minor amount of signal. The star shape is the point-spread function of the imager (i.e., the shape of the volume from which the signal is being received). An image is scanned by moving the object through the sensitive point. This method has produced images from small animals (8,9) and has been used for human imaging by Damadian et al. (10).

The other sensitive-point method was developed by Hinshaw (3) and uses alternating currents applied to three gradient coils to superimpose alternating gradients on the main field. An alternating current applied to a set of gradient coils produces an alternating magnetic field everywhere,

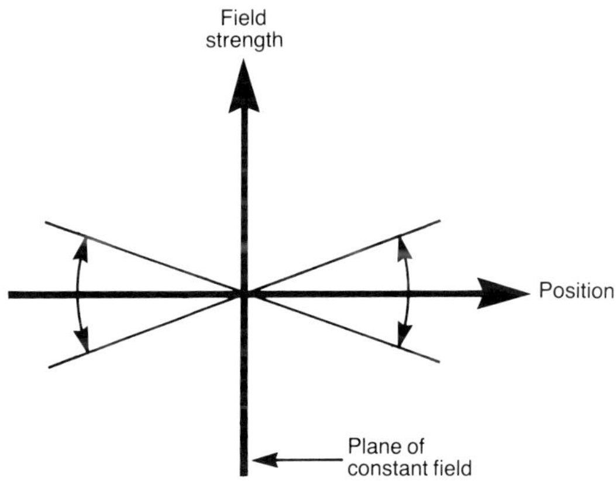

Figure 3.1. Alternating magnetic field resulting from alternating current flowing in a set of gradient coils. The plane at the center of the coils does not experience the alternating field; thus, nuclei in this plane are continuously on resonance.

except in the plane at the center between the coils where the field is unchanged (Fig. 3.1). By applying three alternating currents of unrelated frequencies to three sets of mutually perpendicular coils, three mutually perpendicular planes intersect, so that only the point of intersection has an unchanging field. Areas of the object outside the point of intersection are exposed to the varying magnetic field, and the resonance signals from nuclei outside the sensitive point are frequency modulated by the variations in field. The received signal is thus a combination of the on-resonance signal from the point of intersection and the frequency-modulated signals from the rest of the object. Electronic filtering removes the modulated part of the signal, leaving the signal from the point of intersection.

This point can be moved by varying the ratio of the currents in the coil pairs of each gradient set, allowing electronic scanning of the object. The point-spread function of this method is a cube with rounded edges. The signal observation used with the alternating gradient is the continuous-wave pulse sequence previously described.

The signal dependence on T_1 and T_2 is not the simple, easily separable exponential decay present in other techniques. For the typical range of T_1 and T_2 in tissues, one can approximate the rather complicated exact dependence given in Reference 3 by

$$I - HT_2/T_1, \qquad [11]$$

where H is the hydrogen content.

Image intensities using this technique are not comparable to intensities from other techniques, and obtaining separate T_1 and T_2 images is more of a challenge with this technique.

Line Techniques

As mentioned previously, imaging speed can be improved considerably if all the points along a line are observed simultaneously in the presence of a field gradient along the line. A Fourier transformation is then used to separate the points along the line by correlating resonant frequency with position on the line. Images are made more rapidly, but the techniques require computer processing for the Fourier transformation and more complex signal excitation hardware and software. Line techniques have a lower sensitivity to chemical information than do single-point techniques. However, they are less demanding of magnetic field and field gradient uniformity than are the reconstruction techniques to be described later (1). The many different line-scan techniques (11–29) have developed from two basic sources, Hinshaw's sensitive point, which we have just described, and Garroway et al.'s selective irradiation (4).

Selective Irradiation

A brief description of the general concept of selective irradiation will be helpful because it is important in both linear and planar imaging. Selective irradiation differs from Hinshaw's technique because the whole object is not irradiated with a broad range of radio frequencies. Instead, the range of radio frequencies used is narrow and excites only a restricted volume. Filtering of "background" signal is unnecessary in this technique because the only signal that is observed is the signal from the resonant volume.

Selective irradiation restricts a volume by the following sequence: First, a field gradient is applied across the object. A particular frequency now corresponds to a plane of constant frequency perpendicular to the gradient's direction. While this gradient is present, a low-level radio frequency pulse of relatively long duration (e.g., 10 msec) is applied to the object. The radio frequency pulse is chosen so that its frequency spectrum is a narrow block of frequencies (e.g., a range of 1 kHz). The region in the object whose nuclei resonate at the frequencies contained in the radio frequency pulse will be excited, but the rest of the object is not. In this way, a planar slice through the object can be activated (Fig. 3.2). The slice thickness and position depend on the radio frequency pulse's spectrum and on the strength of the gradient. The position of the plane can be moved by simply changing the frequency of the pulse, thereby activating nuclei at another field strength.

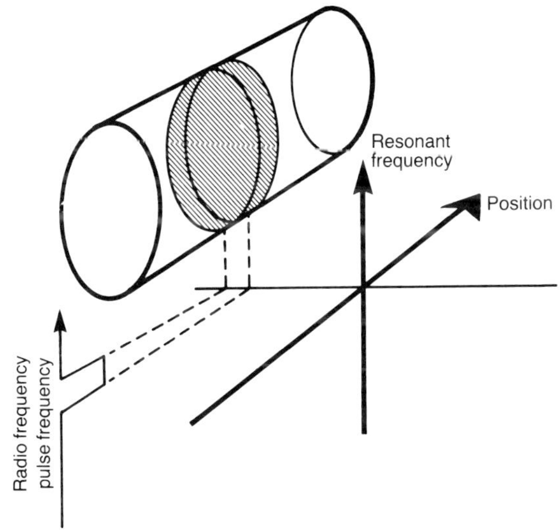

Figure 3.2. Shaded plane is selectively irradiated in the presence of a field gradient. The gradient produces a linear variation of resonant frequency with position. The position and thickness of the plane are defined by the interaction of the radio frequency pulse's frequency spectrum and the object's resonant frequency distribution, as shown by the dashed lines.

Selective Lines

When selective irradiations are used to define a line volume, a sequence of at least two irradiations with different selected directions are used. The type, order, and timing of the irradiations and signal observations determine how the line is defined and the T_1 and T_2 dependence of the signal intensity. The selective irradiation sequence we use in our work (24–28) consists of a 90-degree selective irradiation that defines a plane perpendicular to the y axis. Then a 180-degree selective irradiation inverts the nuclei in a plane perpendicular to the z axis. At a later time, a spin-echo signal will arise only for the nuclei that have experienced both the 90- and 180-degree irradiations, that is, the line volume along the x axis that is the intersection of the two perpendicular planes (Fig. 3.3). Applying an x gradient during the spin echo allows one to use a Fourier transformation to separate the points along the line (Fig. 3.4). The line can be scanned through an object by methodically changing the frequency of either of the selective irradiations, thus exciting planes at different magnetic field strengths. The technique has produced rat images with resolution of 0.5 × 2 × 8.4 mm in 4 min. The effect of motion is, again, to blur or displace individual lines.

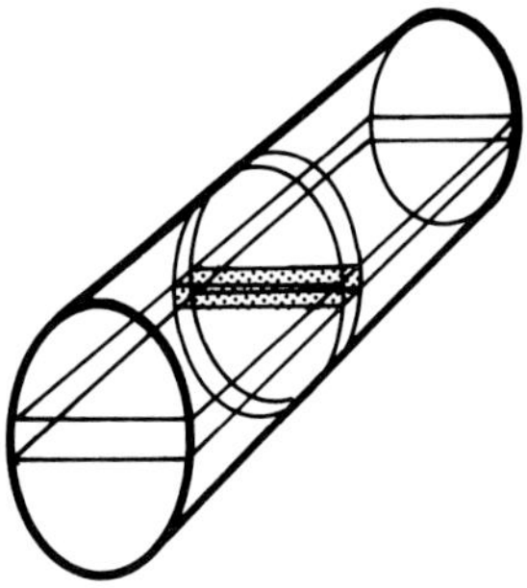

Figure 3.3. The line of intersection of two selectively irradiated planes defines the line that produces the spin-echo signal.

The dependence of signal intensity on T_1 and T_2, for single-compartment relaxation times and no flow, is

$$I = \mathrm{H}[(1 - \exp(-a/T_1)] \exp(-b/T_2), \qquad [12]$$

where a and b are time intervals in the irradiation sequence. The T_2 dependence results from the use of the spin echo for signal observation. Because spin-echo techniques involve observing the signal after a time delay, transverse relaxation processes reduce the signal strength. Techniques that use an FID will not exhibit a T_2 dependence because the signal is observed before transverse relaxation has a major effect.

Other selective irradiation sequences can have a different dependence

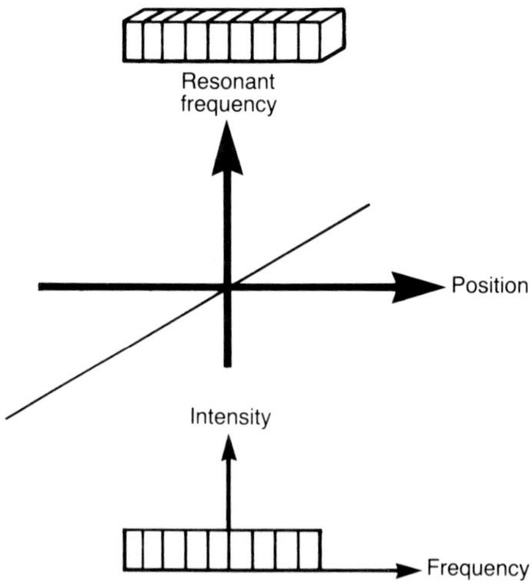

Figure 3.4 A field gradient along the line producing the spin echo spreads the frequencies of the nuclei, so that the frequency spectrum separates the points along the line.

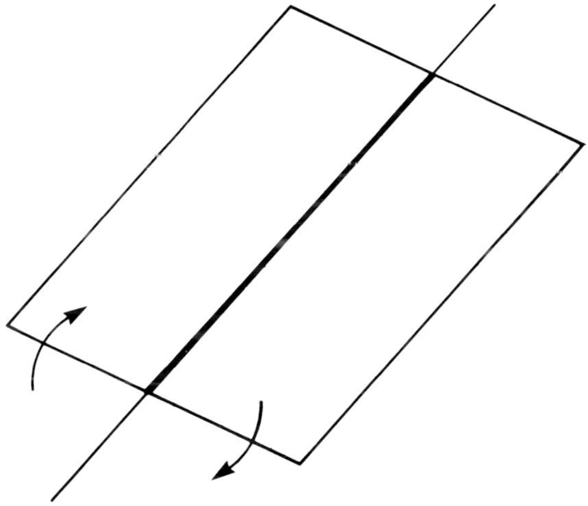

Figure 3.5. Two alternating gradients produce a rotating magnetic field that is constant only along the axis of rotation. Only nuclei along this axis will be constantly on resonance, thus forming a sensitive line.

on T_1. For example, the T_1 dependence of Hutchison et al.'s sequence (29) is

$$I = \mathrm{H}[1 - 2\exp(-a/T_1)]. \qquad \textbf{[13]}$$

Because this technique uses an FID, there is no T_2 dependence. To define a line using this sequence requires the subtraction of two planar signals. The technique has a strong dependence on motion. This characteristic can be seen in the human chest image of Reference 29, where the motion of the heart produces intense areas in the image outside the subject's body. One of the selective irradiation sequences developed by Mansfield et al. (21) has produced human images with a resolution of about 0.5 cm in 40 min, with no obvious motion artifacts.

Sensitive Line

In Hinshaw et al.'s sensitive line (11–13), an extension of their sensitive-point technique, two alternating electric currents of the same frequency but of different phases are applied to two pairs of gradient coils. These two varying gradient fields combine to form a rotating field gradient. Only along the axis of rotation, a linear volume, is the magnetic field constantly on resonance, so this axis is the only region of the object that is observed once all the time-dependent signal has been filtered (Fig. 3.5). Applying a constant gradient along the line allows the use of the Fourier transformation to observe the individual points along the line, by correlating frequency with position on the line. The line can be scanned through an object to produce a planar image by varying the ratios of the currents in the gradient coil pairs. Because the signal is again observed as a continuous wave, its intensity is I = H T_2/T_1. Again, obtaining T_1 and T_2 images poses analytic challenges. This technique has produced images of animals (11–15) and a human head (16) with resolutions of 0.5 and 2 mm, respectively, in times of 2 min. The effect of motion in these images is to blur or displace individual lines.

The sensitive-line techniques are more efficient than single-point techniques and are usually resistant to motion artifacts. Gating can be used to stop motion, as it can in single-point techniques. The authors' technique has been shown to be sensitive to blood flow (27,28), blood moving through the slice having a reduced intensity. With the exception of Hinshaw et al.'s sensitive line, T_1 and T_2 can be investigated by varying timing parameters in the irradiation sequences. The presence of the gradient during the signal observation complicates the derivation of chemical information, a restriction that applies as well to planar and whole-volume techniques.

Planar and Whole-Object Techniques

Planar Volumes

Zeugmatography. Two techniques produce images based on signals arising simultaneously from all points in a plane: zeugmatography, developed by Lauterbur (1), and two-dimensional Fourier transformations, proposed by Kumar et al. (5). Zeugmatography, like x-ray CT, can reconstruct a two-dimensional image from projections of a plane onto a line. Each projection is obtained by exciting all the nuclei in a plane, then observing the resulting signal in the presence of a linear gradient. All the points in each line through the object perpendicular to the gradient are at the same frequency, so each frequency defined by the gradient is equivalent to a projection onto a point of all the points on a line through the plane (Fig. 3.6). Fourier analysis of the NMR signal separates the frequencies along the gradient, providing a set of points that are the projection of the plane onto a line. This process is repeated many times, each time with a different gradient direction (corresponding to rotating the x-ray tube in a CT scan), until a collection of projections is accumulated from which a computer can reconstruct a two-dimensional image of the plane (Fig. 3.7).

To reconstruct an image with $N \times N$ pixels requires at least N projections of N points. The advantage of this technique over sensitive lines is that the NMR signal that arises from the entire plane is much stronger

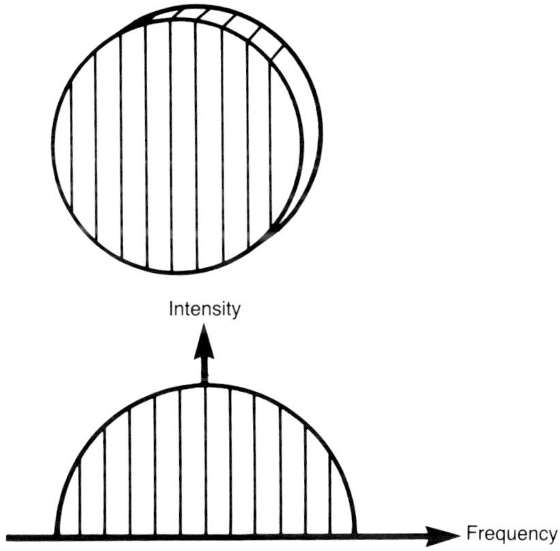

Figure 3.6. Projection of nuclei in a plane onto a line. In the presence of a field gradient, all the nuclei along each vertical line in the plane have the same frequency, so each frequency in the spectrum corresponds to a projection of a line of nuclei onto a single point.

than the signal from a single line and, since the electronic noise is the same in both cases, the final signal to noise of the reconstructed image is better than a sensitive-line image (30). This improvement is counteracted by "reconstruction algorithm noise" (31). Another important, adverse effect on image quality is motion artifacts. Because the image is reconstructed from projections in a manner similar to x-ray CT, the effect of motion should produce artifacts similar to those observed in x-ray CT. This effect will be especially important because NMR imaging time is several minutes, as it was in the older CT scanners. Artifacts due to the beam hardening produced by bone in x-ray CT will be absent in NMR.

Several human images have been generated by reconstruction techniques. Lauterbur has produced projection images of phantoms (32) and, with House, of a rat by this method (33). Head images have been produced by Hawkes et al. (34,35) using a projection technique. In their technique, a plane through the head is defined by a single alternating gradient used in the same way that the two or three alternating gradients were used by Hinshaw et al. to define a line or a point. These images were taken in 2 min and have a resolution of about 2 × 2 × 10 mm. Contrast in the images is sufficient to distinguish grey from white matter in the brain. No motion artifacts are evident in these images, but the head can easily be

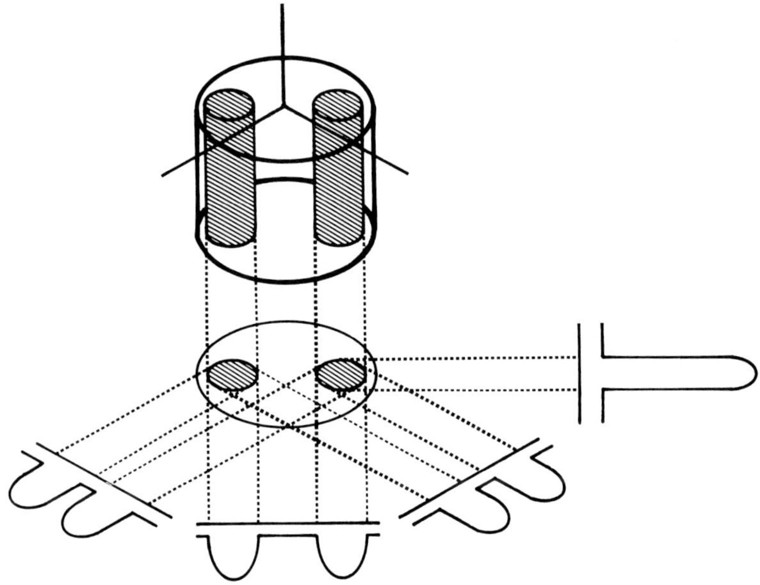

Figure 3.7. By acquiring projections of an object using many different gradient directions, a two-dimensional image of a plane can be reconstructed. From Lauterbur (1) by permission of *Nature (London)*.

held still for 2 min. Researchers still question the effect of motion on abdominal images. Central-point artifacts are noticeable in some images. Because the plane is defined by an alternating gradient and the signal is observed as a continuous wave, the image intensity is again $I = \mathrm{H}\, T_2/T_1$. A selective irradiation can be used to excite nuclei in the plane, and an FID or spin echo (or both) can be used to observe the signal, allowing T_1 and T_2 to be measured directly from the exponential behavior of the signal.

Two-dimensional Fourier Transformation. Two-dimensional Fourier transformation imaging projects the entire plane onto a line, as in reconstruction techniques. Instead of progressively changing the direction of the line (i.e., changing the direction of the gradient across the plane), the strength of a gradient perpendicular to the line of projection is varied to produce phase encoding of position along the perpendicular dimension (Fig. 3.8). As the strength of the perpendicular gradient is increased, phase encoding is progressively increased, simulating a frequency distribution, so that the encoded phase can be decoded using a Fourier transformation.

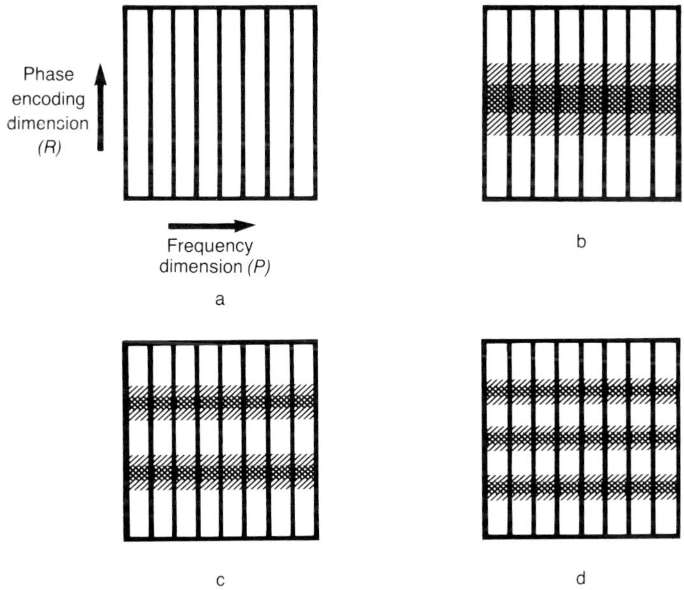

Figure 3.8. Two-dimensional Fourier transformation projections of a plane with progressively increasing phase encoding. (a) Zero-phase encoding acquires uniform signal from all nuclei in each vertical line as single points in the frequency spectrum. (b) First level of phase encoding. When the gradient (*R*) is changed, the phase of nuclei in the center strip is 180 degrees off the phase of nuclei producing positive signals; thus, their signal subtracts from the total signal. The gray areas denote positions in the gradient field where nuclei experience a phase shift of more or less than 180 degrees with respect to the phase of the nuclei producing positive signals. (c) Second level of phase encoding is twice the amount of the first level and produces two bands of nuclei whose signals subtract rather than add to the projection. (d) Third level of phase encoding is three times the amount of the first level and produces three bands of negative signals. The phase encoding is increased through as many levels as there are lines to be resolved in the image. The collection of increasingly phase-encoded projections is then Fourier transformed to obtain the image.

Perhaps we should elaborate slightly. In zeugmatography, a single gradient is rotated electronically around the plane. Lines in the plane are projected onto points on a line from about as many different angles as there are lines in the picture (Fig. 3.7). However, in two-dimensional Fourier transformations, the line of projection is fixed. A second gradient, whose axis is at right angles (R) to the gradient that defines the projected line (P), provides the additional information needed to create a two-dimensional image. When the signal is observed, the frequency spectrum occasioned by gradient P equals a projection onto a set of points of all the points in the lines at right angles to the direction of P (Fig. 3.8). If the second gradient is zero, the signal will be Fourier transformed, as in zeugmatography, into one projection of the plane. If the orthogonal gradient (R) is applied briefly before the P gradient, nuclei will begin to precess at frequencies determined by their position in the R-gradient field. When the R gradient is removed and the P gradient applied, although all the nuclei in a line will resonate at the same frequency, points on a line where nuclei precessed at higher frequencies when exposed to gradient R will have a different phase from other points on the same line (Fig. 3.8). Phase can, therefore, give information about spin distribution.

Again, to produce an image of $N \times N$ pixels, N projections of N points must be acquired. Instead of changing the direction of the projections, the amount of the phase encoded by the gradient is increased for each of the projections. The corresponding points in each of the N increasingly phase-encoded projections are Fourier transformed to produce the final image. Gradients in either the main magnetic field or in the radio frequency magnetic field used to excite the nuclei are capable of producing the phase encoding. A radio frequency gradient is created by reversing current direction in one half of a radio frequency coil pair. Hoult has produced a phantom object image using the radio frequency field gradient (36).

Edelstein et al. have produced human images using a gradient of the main field for phase encoding (37). This two-dimensional Fourier transformation technique uses a selective irradiation to select the plane to be imaged. T_1 images can be produced by combining two images taken with different initial radio frequency pulses. The body images done by Edelstein et al. have a resolution of 7.5 × 7.5 × 18.5 mm and require 128 sec. No motion artifacts are noted. The interventricular septum of a live heart is clearly visible in the T_1 image (Chapter 6).

Because the two-dimensional Fourier transformation technique receives the signal from the entire plane, it has a signal-to-noise advantage over sensitive-line techniques. Its response to motion is not yet known, but it is not expected to be like that of CT because of the nature of the projection. However, because the intensity of each pixel depends on every point sampled to produce each projection from which the image is reconstructed, motion artifacts could spread through many lines in the object and be more objectionable than motion artifacts in sensitive-line techniques.

In summary, the two planar imaging techniques just described require a computer to perform the reconstruction. The algorithms for zeugmatography are more complex because the fast Fourier transformation, which is a standard computer algorithm, is used for two-dimensional Fourier transformations. The magnetic field and field gradients have to be very uniform to produce good projections for zeugmatography. Because the planar signal is large, one has a signal-to-noise advantage over sensitive-point or line techniques, but this advantage places greater demands on the instruments because a greater dynamic range is required of the receiver. The reconstructions require signals from all points in the object, so that each point cannot be uniquely determined. Reconstructed images require a set number of projections; therefore, the number of picture elements cannot be varied during imaging, unlike the situation with single-point and line techniques, in which an image itself can be restricted to a specific area of interest. In reconstruction techniques, one cannot focus signal acquisition on a region of interest; one can only magnify a region in the completed image. As in the sensitive-line techniques, the presence of a gradient complicates the determination of chemical information. Using phantoms, chemical images have been made (32,38). Gating can be applied in the same way as for selective lines.

Whole Objects

It is possible to extend the simultaneous observation of the signal to all the points in an object, not just to the points in a plane. This results in the reconstruction of an entire three-dimensional object and can be done by zeugmatography or by three-dimensional Fourier transformation. As far as signal acquisition is concerned, both are extensions of the two-dimensional procedures. For zeugmatography, the entire object is projected onto a line, and the direction of the line is rotated in three dimensions, so that N^3 pixels can be determined from N point projections taken over N^2 unique directions. For the three-dimensional Fourier transformation, the object is projected onto a line, and phase encoding in the two directions perpendicular to the line of projection is done by systematically varying the gradients in these two directions, so that N^2 projections, each with a unique phase encoding, of N points can be reconstructed into N^3 pixels. Only the zeugmatographic technique has been implemented.

Because the signal arises from the entire object, it is stronger than the signal from a single plane; therefore, these techniques have a signal-to-noise advantage over planar and linear imaging techniques. For producing multiplanar images, it is not clear that this advantage exceeds that which would be achieved by using a planar technique that allows the sequential selective irradiation of many planes, the signal from one plane being observed while the other planes are returning to equilibrium. An advantage of whole-object techniques is that all pixels are the same size, so image slices have the same thickness as they have resolution. Because most other techniques are slice oriented, they have a slice thickness larger than the resolution within the plane. In zeugmatography, because the angles of projection are all equal, the pixels are cubed elements. The computer can also be used to select any view of the three-dimensional collection of pixels.

The whole-object imaging techniques have increased technical requirements and enhanced performance characteristics that are direct extensions of their planar versions. The problem of storing and reconstructing N^3 pixels will require substantial computer resources. Effective approaches to the reconstructions are being developed (39–43), but their artifacts are as yet unknown. Lauterbur has used three-dimensional zeugmatography to produce images of human brains (44).

PERFORMANCE PARAMETERS

Speed

Several basic performance characteristics of imaging systems determine areas of investigation for which a system is suited. For example, the speed of ultrasonic imaging allows the viewing of real-time images of moving organs, while slower imaging systems require gating. The time required to produce an NMR image is slow, being on the order of minutes. At present, the best NMR images of the human head require about 2 min (34,35). Whole-body images will take somewhat longer.

Long NMR imaging times are an intrinsic, rather than technical, restriction. The resonant nuclei emit very-low-level radio signals. When these signals are picked up by a coil of wire surrounding the patient and used to form the image, the coil also picks up thermally generated radio signals from the patient, which produce noise in the image. To reduce the effect of this noise on the image, a signal is elicited from each element of the image many times and averaged. This averaging reduces the noise in relation to the strength of the desired signal. However, the repeated acquisition of the signal from each element increases the time required to produce the image. At present, it seems unlikely that improvements of more than a factor of 10 can be made in this speed.

Motion

The relatively slow imaging speed requires that another characteristic of NMR imaging be considered, namely, sensitivity to motion. In many imaging techniques, blood flow reduces blood's signal intensity, and this effect may be used to advantage to measure blood flow or to detect obstructions in blood vessels. NMR imaging techniques that reconstruct images from projections should have motion artifacts that are similar to those encountered in x-ray CT, although these artifacts may be less serious. A selective irradiation technique, in which a line is defined by subtracting two planar signals obtained at different times, demonstrates extreme artifacts due to heart motion (29). Techniques in which points or lines are directly defined by the NMR process will only exhibit blurring or offsets of the particular line due to motion. This effect has been observed in several of the line scan techniques (12,25).

NMR, like nuclear medicine and x-ray CT, will require gating to study the motion of the heart. Electronic control of the imaging process simplifies the gating process. The time required to acquire the signal used in the averaging is about 50 msec, sufficiently short to be compatible with heart motions.

Signal to Noise

In NMR, the signal-to-noise ratio (S/N) in the instrument depends on a variety of factors, including strength of the magnetic field, resolution volume, total volume of the system, imaging technique used, and imaging time. However, it is important to realize that the value of S/N in the instrument is not equivalent to S/N in the final image.

S/N in the final image depends on both the chosen spatial resolution and the imaging time. The spatial resolution is given by the volume from which signals will be received. Because noise is determined mostly by the instrument, larger volumes produce larger signals and improved S/N. On the other hand, as the size of the resolution volume increases with respect to object size, blurring will decrease constrast in the output image in a predictable manner (45,46). Thus, when imaging a small-feature object with a relatively large resolution volume, although S/N increases for the instrument, an adverse effect is noted on the S/N of the output image.

In this context, we note that FID signal readout provides stronger signals than do pulse echos, thus yielding improved S/N for the instrument. On the other hand, T_2 information, and the contrast it provides, is lost. Thus, the lower signal in the pulse echo (typically 40% to 60% of the signal level of the FID) may actually provide a higher S/N in the output image by enhancing contrast between tissues with different T_2 relaxation times.

CONCLUSIONS

The relative advantages and disadvantages of different imaging techniques and readout modes (FID versus pulse echo) are poorly understood.

Theoretic evaluation of S/N has little meaning unless done in a manner consistent with the reality of object characteristics and the constraints imposed by hardware limitations. The "ideal" imaging technique is a worthy grail to seek. We will have to wait until clinical systems start to undergo evaluation under routine conditions to even begin the search. Maybe an ideal technique does not exist. In x-ray CT, second-, third-, and fourth-generation scanners are produced that yield images that belie the implications of the terminology used to describe them. Sound engineering, although lacking the glamor of first principles analysis, is likely to be the determinant that will grade the utility of different NMR imagers.

REFERENCES

1. Lauterbur PC: Image formation by induced local interactions: Examples employing nuclear magnetic resonance. *Nature (London)* 242:190–191, 1973.
2. Damadian R: U.S. Patent 3,789,832, filed March 17, 1972.
3. Hinshaw WS: Image formation by nuclear magnetic resonance: The sensitive-point method. *J Appl Phys* 47:3709–3721, 1976.
4. Garroway AN, Grannell PK, Mansfield P: Image formation in NMR by a selective irradiative process. *J Phys C* 7:457–462, 1974.
5. Kumar A, Welti D, Ernst R: NMR Fourier zeugmatography. *J Magn Reson* 18:69–85, 1975.
6. Hoult DI: NMR imaging. Rotating frame selective pulses. *J Magn Reson* 38:369–374, 1980.
7. Lauterbur PC, Lai C: Unpublished work.
8. Damadian R, Minkoff L, Goldsmith M, et al: Field focusing nuclear magnetic resonance (FONAR): Visualization of a tumor in a live animal. *Science* 194:1430–1431, 1976.
9. Crooks L, Grover T, Kaufman L, et al: Tomographic imaging with nuclear magnetic resonance. *Invest Radiol* 13:63–66, 1978.
10. Damadian R, Goldsmith M, Minkoff L: NMR in cancer: XX. FONAR scans of patients with cancer. *Physiol Chem Phys* 10:285–286, 1978.
11. Andrew ER, Bottomley PA, Hinshaw WS, et al: NMR images by the multiple sensitive point method: Application to larger biological systems. *Phys Med Biol* 22:971–974, 1977.
12. Hinshaw WS, Bottomley PA, Holland GN: Radiographic thin-section image of the human wrist by nuclear magnetic resonance. *Nature (London)* 270:722–723, 1977.
13. Hinshaw WS, Andrew ER, Bottomley PA, et al: Display of cross sectional anatomy by nuclear magnetic resonance imaging. *Br J Radiol* 51:273–280, 1978.
14. Hinshaw WS, Andrew ER, Bottomley PA, et al: An in vivo study of the fore-arm and hand by thin section NMR imaging. *Br J Radiol* 52:36–43, 1979.
15. Bottomley PA: In vivo tumor discrimination in a rat by proton nuclear magnetic resonance imaging. *Cancer Res* 39:468–470, 1979.
16. Holland GN, Moore WS, Hawkes RC: Nuclear magnetic resonance tomography of the brain. *J Comput Assist Tomogr* 4:1–3, 1980.

17. Mansfield P, Maudsley AA, Baines T: Fast scan proton density imaging by NMR. *J Phys E* 9:271–278, 1976.
18. Mansfield P, Maudsley AA: Line scan proton spin imaging in biological structures by NMR. *Phys Med Biol* 21:847–852, 1976.
19. Mansfield P, Maudsley AA: Medical imaging by NMR. *Br J Radiol* 50:188–194, 1977.
20. Pykett IL, Mansfield P: A line scan image study of a tumorous rat leg by NMR. *Phys Med Biol* 23:961–965, 1978.
21. Mansfield P, Pykett IL, Morris PG, et al: Human whole body line-scan imaging by NMR. *Br J Radiol* 51:921–922, 1978.
22. Mansfield P, Morris PG, Ordidge R, et al: Carcinoma of the breast by nuclear magnetic resonance (NMR). *Br J Radiol* 52:242–243, 1979.
23. Morris P, Mansfield P, Pykett I, et al: Human whole-body line scan imaging by nuclear magnetic resonance. *IEEE Trans Nucl Sci* NS-26:2817–2820, 1979.
24. Crooks LE, Hoenninger JC, Arakawa M, et al: Tomography of hydrogen with NMR, and the potential for imaging other body constituents. *INSERM* 88:19–34, 1979.
25. Crooks LE, Hoenninger JC, Arakawa M, et al: Tomography of hydrogen with NMR and the potential for imaging other body constituents. *SPIE J* 206:120–128, 1979.
26. Crooks LE: Selective irradiation line scan techniques for NMR imaging. *IEEE Trans Nucl Sci* NS-27:1239–1244, 1980.
27. Hansen G, Crooks LE, Davis PL, et al: In vivo imaging of the rat anatomy with nuclear magnetic resonance. *Radiology* 136:695–700, 1980.
28. Crooks LE, Hoenninger JC, Arakawa M, et al: Tomography of hydrogen with nuclear magnetic resonance. *Radiology* 136:701–706, 1980.
29. Hutchison J, Edelstein W, Johnson G: A whole-body NMR imaging machine. *J Phys E* 13:903–1030, 1980.
30. Brunner P, Ernst RR: Sensitivity and performance time in NMR imaging. *J Magn Reson* 33:83–106, 1979.
31. Budinger TF, Derenzo SE, Gullberg GT, et al: Emission computer assisted tomography with a single-photon and positron annihilation photon emitters. *J Comput Assist Tomogr* 1:131–145, 1977.
32. Lauterbur PC, Kramer DM, House WV, et al: Zeugmatographic high resolution NMR spectroscopy: Images of chemical inhomogeneity with macroscopic objects. *Am Chem Soc J* 97:6866–6868, 1975.
33. House WV: Introduction to the principles of NMR. *IEEE Trans Nucl Sci* NS-27:1220–1226, 1980.
34. Holland GN, Moore WS, Hawkes RC: Nuclear magnetic resonance tomography of the brain: Coronal and saggital sections. *J Comput Assist Tomogr* 4:429–433, 1980. *J Comput Assist Tomogr* 4:577–586, 1980.
35. Hawkes R, Holland G, Moore W, et al: Nuclear magnetic resonance (NMR) tomography of the brain: A preliminary clinical assessment with demonstration of pathology. *J Comput Assist Tomogr* 4:577–586, 1980
36. Hoult DI: Rotating frame zeugmatography. Paper presented at 19th Experimental NMR Conference, Blackburg, Virginia, 1978.
37. Edelstein W, Hutchison J, Johnson G, et al: Spin warp NMR imaging and applications to human whole-body imaging. *Phys Med Biol* 25:751–756, 1980.
38. Bendel P, Lai C, Lauterbur PC: ^{31}P spectroscopic zeugmatography of phosphorus metabolites. *J Magn Reson* 38:343–356, 1980.

39. Marr R, Chen C, Lauterbur PC: On two approaches to 3D reconstruction in NMR zeugmatography. Brookhaven National Laboratory, Report BNL-28041/AMD-857, February 1980.
40. Lai C, Lauterbur PC: True three-dimensional image reconstruction by nuclear magnetic resonance zeugmatography. In press.
41. Lai C, Lauterbur PC: A gradient control device for complete three-dimensional nuclear magnetic resonance zeugmatographic imaging. *J Phys E* 13:747–751, 1980.
42. Shepp LA: Computerized tomography and nuclear magnetic resonance. *J Comput Assist Tomogr* 4:94–107, 1980.
43. Lauterbur PC, Lai CM: Zeugmatography by reconstruction from projections. *IEEE Trans Nucl Sci* NS-27:1227–1231, 1980.
44. Lauterbur PC: Unpublished work.
45. Shosa DW, Kaufman L: Methods for evaluation of diagnostic imaging instrumentation. *Phys Med Biol* 26:101–112, 1981.
46. Kaufman L, Shosa DW: Generalized methodology for the comparison of diagnostic imaging instrumentation. *AFIPS Nat Comput Conf Expo Conf Proc* 49:445–451, 1980.

4

Hardware for NMR Imaging

Leon Kaufman, Ph.D.

Lawrence E. Crooks, Ph.D.

GENERAL CONSIDERATIONS

Although different imaging techniques impose considerably different requirements on the specifications of the NMR imager hardware, the major components are common to all techniques. Figure 4.1 shows the block diagram of the typical NMR imager: A magnet is used to impose a strong, uniform magnetic field. Gradient coils modulate this uniform field in a manner dictated by the imaging technique in use. To induce resonance, a transmitter delivers radio frequency power through an antenna. The antenna also picks up the NMR signals, while a receiver, connected to the antenna, delivers the suitably amplified and filtered signals to digitizing circuitry. Finally, a computer-based system is provided for control, data-acquisition, processing, and display functions.

MAGNETS

A macroscopic magnetic field can be generated by either permanent magnets (Fig. 4.2) or electric current, which produces a magnetic field around the conductor, as shown in Figure 4.3. These two methods can be combined: Current-carrying conductors are often used to induce a field in a magnetically susceptible material, such as iron (Fig. 4.4). In general, conductors can be resistive (e.g., copper or aluminum wires), in which case a power supply provides a driving voltage continuously, or they may be superconducting (i.e., cooled to extremely low temperatures so there is no electrical resistance in the conductor), in which case once a current has been established in the conductor, the power supply can be disconnected.

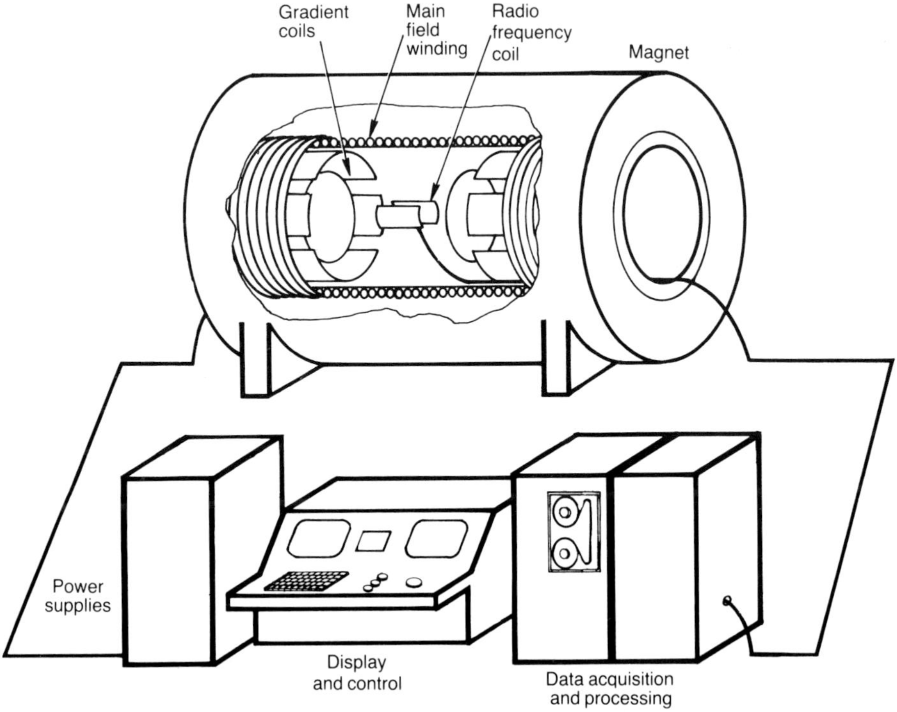

Figure 4.1. Block diagram of an NMR imaging system.

These three methods of generating a magnetic field, reflected in the basic iron-core, resistive air-core, and superconducting air-core magnet designs, have all been used for NMR imaging. Choices by individual research groups reflect a balance between their technical needs for strength and uniformity of field and practical considerations of magnet size, weight, cost, and availability.

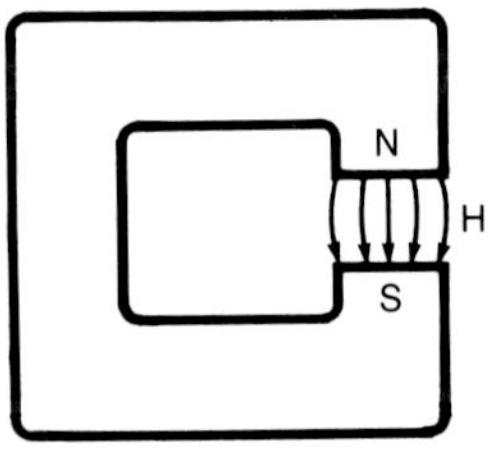

Figure 4.2. A magnetic field can be generated in the gap of a permanent magnet.

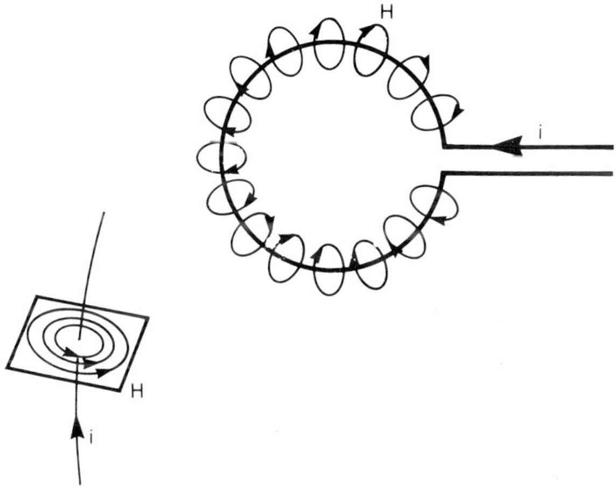

Figure 4.3. A magnetic field H is generated around a wire carrying a current i.

Iron-Core Magnets

These magnets consist of a C- or H-shaped core of soft iron (in which it is easy to establish a magnetic field) surrounded by resistive coils through which a direct current flows, establishing the field (Figs. 4.4. and 4.5). Magnetic flux lines are almost completely confined to the iron, so comparatively small currents can produce relatively high and uniform fields. The uniformity is established by careful machining and polishing of the iron pole tips; it is on the order of 45 ppm in a 4-cm region for pole tips 30 cm in diameter and 10 cm apart. When large apertures are needed, the weight and power demands of an iron-core magnet may become prohibitive. But when small apertures, 10 cm diameter or less, are adequate, these magnets offer several advantages over alternative designs: The iron mass provides a compact and reliable way to obtain uniformity; magnet

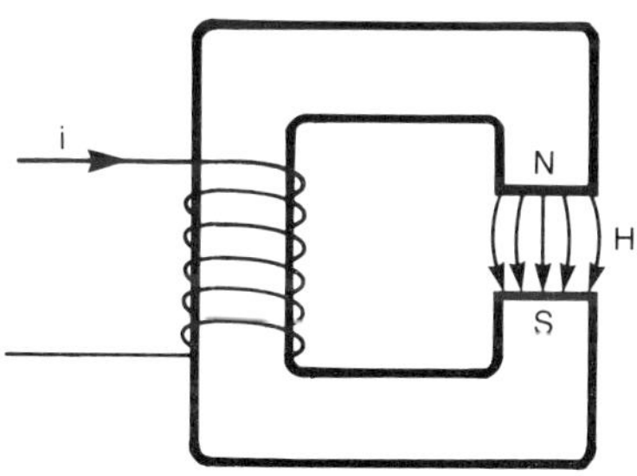

Figure 4.4. A current can induce magnetization in "soft" iron, which produces a magnetic field in the gap of an iron-core magnet. This configuration is called a "C" magnet.

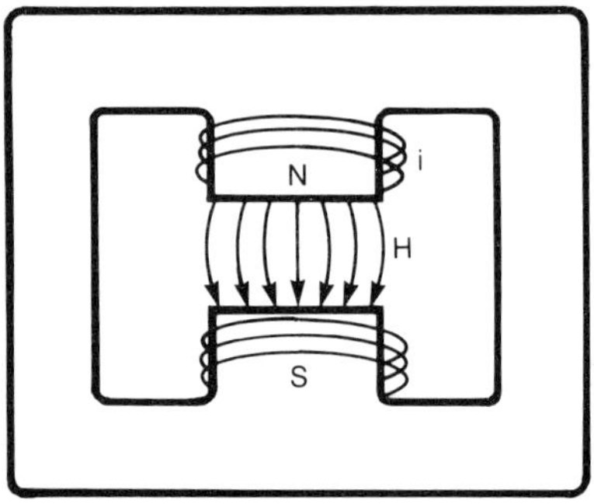

Figure 4.5. In an "H" magnet, two coils magnetize the iron to produce a strong, uniform field in the gap.

costs are low; commercial availability is good. For a pole-tip aperture of 10 cm, fields of 7 kG can be obtained with 20-kW power supplies. A typical iron-core magnet is shown in Figure 4.6.

Resistive-Coil, Air-Core Magnets

If magnetic materials are not used to shape the field, the configuration of the current-carrying conductors becomes the determining factor in establishing field uniformity. In theory, a uniform current density over the surface of a sphere will produce a perfectly uniform magnetic field within the sphere (Fig. 4.7). Although this ideal spherical configuration is hard to realize in practice, it can be approximated by simpler configurations. The simplest such approximation is the Helmholtz coil: two circular coils of equal radius, with a common axis, separated by a distance equal to the radius (Fig. 4.8). The current flows in the same direction in both coils. However, the Helmholtz coil does not provide good field uniformity over a large volume. A configuration that produces a more uniform field consists of two pairs of coils, with a common axis, each placed on the surface of a hypothetic sphere (Fig. 4.9). Uniformities of 1 part in 10^3 over a 50-cm diameter, or of 1 part in 10^4 over a 35-cm diameter, can be achieved with a four-coil configuration.

In a resistive magnet, power considerations (both production and dissipation) limit the maximal fields that can be reached to 1-1.5 kG. Power needs increase as the square of the magnetic field strength (if conductor temperature is kept constant by adequate cooling). Because the four-coil configuration is a very efficient way to produce a uniform magnetic field for a given amount of power, it has become the favorite design for resistive magnets in human-sized NMR imagers. For a 1.2-kG field, power consumption is 20 kW. A typical resistive-coil, air-core magnet is shown in Figure 4.10.

Figure 4.6. "H"-type iron-core magnet with a 30-cm pole tip and 10-cm gap. Coils are contained in the dark disks. Elliptic opening is the aperture of the radio frequency coil.

Figure 4.7. Current-carrying conductors (small circles) on the surface of a sphere produce a uniform magnetic field inside it (arrows). This configuration is difficult to achieve in practice.

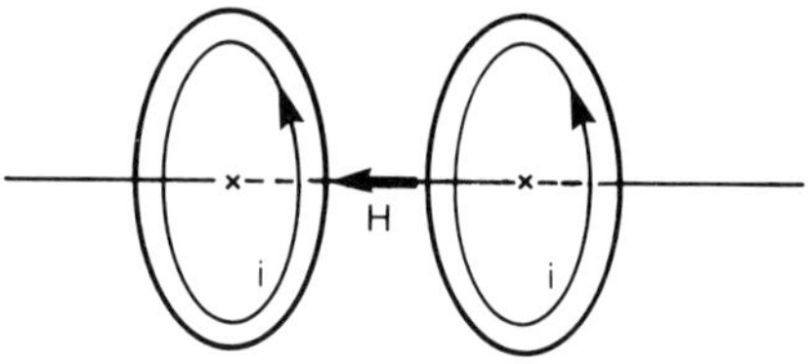

Figure 4.8. An approximation to the spherical current distribution is given by the Helmholtz coil pair.

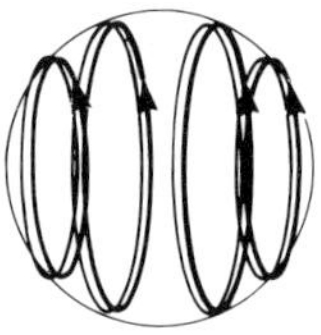

Figure 4.9. Two sets of coils on the surface of a sphere provide a better approximation of the ideal current distribution (Fig. 4.7) than does the Helmholtz pair (Fig. 4.8).

Figure 4.10. Resistive air-core magnet with a 40-cm useful aperture. Small and large sets of coils are evident in this front view. The aperture is occupied by calibration equipment.

For human-size NMR imagers, resistive magnets offer the lowest cost alternative. They are also comparatively easy to manufacture, giving them ready availability. In many cases, these advantages override other considerations, such as their comparatively low field, instabilities due to power supply regulation, and high electric power consumption.

Superconducting Magnets

When field strengths higher than the 1-1.5 kG obtainable with resistive magnets are desired, superconducting air-core magnets presently offer the most practical alternative.

Superconducting magnet design is dominated by the need to provide a cryogenic environment for the superconductor, typically filaments of niobium-titanium (NbTi) embedded in a copper matrix (Fig. 4.11). The conductor is partially submerged in liquid helium, which, in turn, is surrounded by a vacuum, a layer of liquid nitrogen, another vacuum, and insulation. The copper matrix for the NbTi protects the system from destruction in case of an accidental quench (i.e., a loss of superconductivity). When running at superconducting temperatures, the copper serves as a heat conductor, although not an electrical conductor, while the NbTi, which at 4°K has no resistance, conducts the current. However, if the temperature begins to rise, due either to the absence of helium or the loss of vacuum integrity, the copper quickly becomes the less resistive of the two conductors and safely dissipates the current previously confined to the NbTi, thus preventing a "melt-down" of the NbTi.

Although, in principle, the four-coil design can be extended to superconducting magnets, cryostat design of a four-coil superconducting magnet is difficult. The strong forces acting on each coil in a four-coil design require that strong mechanical supports be provided for the coils, whereas the need for good thermal insulation requires that mechanical supports be as thin as possible to reduce conductive heating. Consequently, because they can be suspended as a single unit, solenoidal configurations have been used in human-size superconducting magnets.

A solenoid, too, can be thought of as approximating a section of a sphere (Fig. 4.12). This approximation leads to increasingly better field uniformity as the solenoid increases in length and in diameter; thus, the basic limitation becomes practical acceptable size. A superconducting magnet of the type shown in Figure 4.13, with an outside diameter of 1.5 m and a length of 2 m, can have a field uniformity of 1 part in 10^4 over a 40-cm diameter region. This magnet operates at a magnetic field strength of 3.5 kG. The decay rate of this field is on the order of 1 ppm/hr, thus providing an extremely stable magnetic environment.

Figure 4.11. Superconducting filaments (dark material) are embedded in a copper matrix (light-color material) to form wire for superconducting magnets (Courtesy of Intermagnetics General Corporation).

Superconducting magnets can offer high, uniform, and stable magnetic fields, but they are more expensive than resistive magnets to construct. Liquid helium costs offset the power consumption advantage of superconductors.

GRADIENT COILS

As described elsewhere in this monograph, the NMR imaging process requires spatially variant magnetic fields, that is, gradients. Because gradient coils merely modify an existing magnetic field, their magnetic fields

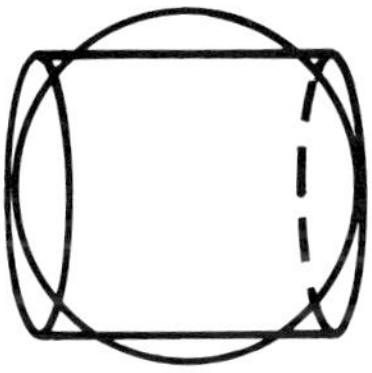

Figure 4.12. A sphere can be approximated by a solenoid, which is a simpler configuration to produce.

need be neither as strong nor as uniform as the main field. However, the presence of the main magnetic field does impose certain design contraints. Gradients must be stronger than the nonuniformity of the main field; otherwise, the uncontrolled nonuniformity will be the dominant factor. Thus, for any given magnetic field, a minimum value for the gradient field is established. Gradient fields are approximately two orders of magnitude weaker than the main field: A typical field gradient for an NMR imager is on the order of 1 G/cm. If a linear gradient is desired over a 30-cm region, the fields must vary from -15 to $+15$ G in that region. Gradient coils must also be designed to switch on and off rapidly (typically within 1 msec) due to the requirement of many imaging methods that the strength or direction

Figure 4.13. Superconducting solenoidal magnet is inside the light-color vacuum can. Dark insert supports the resistive gradient coils.

(or both) of field gradients be varied during the imaging process. This switching generates variable forces on the coils, necessitating careful design to minimize their motion.

All imaging techniques require three independent field gradients. To avoid appreciable distortion in response, gradient configuration must be such as to create a uniform change in the magnetic field over the region of interest. The simplest way of creating a gradient, flowing a current through a wire (Fig. 4.3), creates a nonlinear gradient. Therefore, a better configuration consists of two parallel wires, with currents flowing in opposite directions. In practice, both the Helmholtz coil and a four-coil configuration can be used to generate a highly linear gradient by reversing the current, which for uniform fields flows in the same direction in both coils, in one of the halves of the system. Different imagers use parallel wires, Helmholtz pairs, and four-coil configurations to create each field gradient. Because of spatial limitations inside the magnets, coils are often distorted from their ideal shapes.

ELECTRONICS FOR RADIO FREQUENCY TRANSMISSION AND RECEPTION

Radio Frequency Coil Configuration

Excitation and observation of NMR requires an antenna that couples radio frequency energy to the resonating nuclei. This antenna consists of a copper coil surrounding the subject that is tuned to center its radio frequency on the resonant frequency of the nuclei in the main magnetic field.

The shape of the tuned radio frequency coil is determined by the type of magnet chosen to produce the main magnetic field because the radio frequency magnetic field produced by that coil must be perpendicular to the main field. When imaging human beings, there are only two practical orientations of the main field with respect to the long axis of the body. The main field can be along the body's axis, the case for a person lying in a solenoidal or a four-coil magnet. The radio frequency coil in such cases is the "saddle" shape shown in Figure 4.14a with the radio frequency field pointing upward. The other possible alignment is a main field perpendicular to the body's axis, the case for a person lying between the pole tips of an iron-core magnet or a turned four-coil, air-core magnet. In this instance, the radio frequency coil would be the solenoid shown in Figure 4.14b, with the radio frequency field along the long axis of the person.

Uniformity of the radio frequency magnetic field of the coil will affect

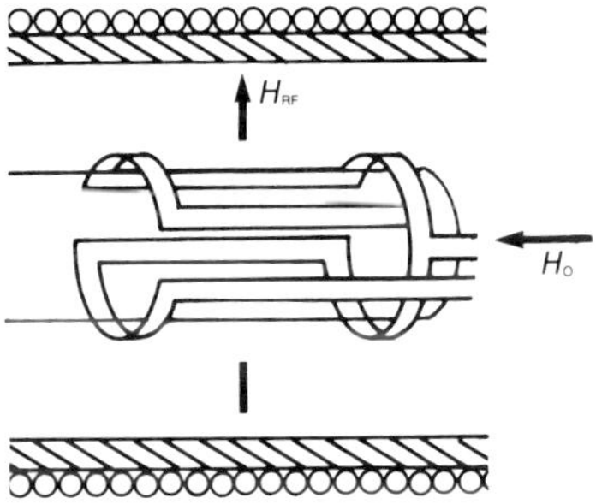

Figure 4.14a. Saddle-shaped radio frequency coil used in solenoidal magnets.

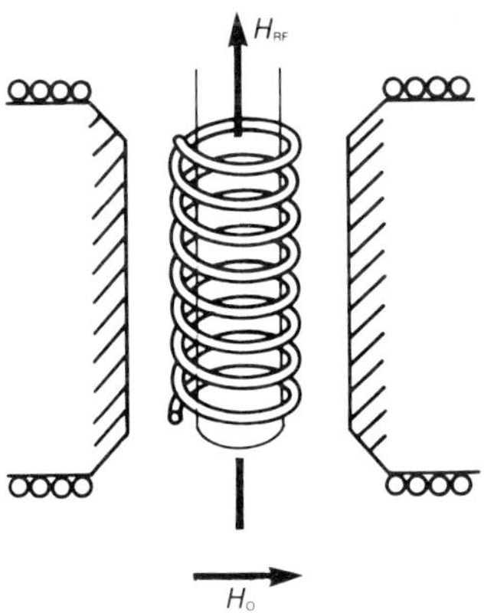

Figure 4.14b. Solenoidal radio frequency coil typically used in iron-core magnets.

the uniformity of the image. Nuclei in the image volume close to the coil wires couple more strongly to the coil and appear brighter than do areas farther from the coil. By imaging a uniform phantom, a flood source correction can be used that eliminates this nonuniformity in the final image.

Most noise in the image is due to electronic noise in the radio frequency coil. Noise also can come from external sources that radiate radio frequencies (the computer is a good example). However, sufficient radio frequency shielding can eliminate external interference. The basic limit on coil noise is electrical resistance or, more exactly, the thermal noise created by electrical resistance. There are two components of the coil's resistance, the most evident being the resistance of the copper coil winding. In addition, however, the resistance of the object being imaged also affects coil resistance because the interaction between the resistance of the body and the electric and magnetic fields of the coil manifests itself as added coil resistance. Tissue resistance, due to the presence of conducting ions, produces more noise than does the resistance of the copper coil windings and, at present, the effect of the body's resistance on the radio frequency coil seems the basic limit on signal noise.

Radio Frequency Transmission

To excite the nuclei in an object, a pulse of radio frequency energy must be applied to the object. A current flowing in the radio frequency coil provides the radio frequency magnetic field needed to excite the nuclei. The current is provided by amplifying a low-level voltage generated in a computer-controlled frequency synthesizer. The current is then established in the coil by applying the radio frequency amplifier output to the coil. Because the radio frequency coil is used for both transmission and reception, it is switched to the output of the radio frequency amplifier only when the nuclei are to be excited. After excitation, the radio frequency coil is switched from the amplifier to the NMR signal receiver.

The strength and duration of the radio frequency pulse varies with the volume selection technique being used. If the entire object is to be excited, the radio frequency amplifier is driven to full output for a short time, for example, 10kW for 50 μsec. If a selected volume is to be excited, the low-level radio frequency voltage will be shaped to have a narrow band spectrum, and the radio frequency amplifier's output at the peak of the shaped pulse will be about 1 kW; the pulse will last a relatively long time, about 5 msec.

NMR Reception

During reception, the radio frequency signal emitted by the nuclei is picked up by the radio frequency coil, amplified by the receiver, and then mixed with the frequency synthesizer output to produce an audio frequency signal that contains the same spectral information as the radio frequency signal. The audio frequency signal is digitized and stored in the computer for processing into an image.

The NMR signal receiver has similar organization and sensitivity to that of a short-wave radio. However, unlike a short-wave radio, which can receive a broad range of frequencies, the NMR signal receiver is designed to be optimized at only the resonance frequency of the nuclei being imaged. The receiver, with a sensitivity to 1 μV, amplifies by 10^6. The extreme sensitivity of the receiver requires not only that the coil be disconnected from the receiver during transmission but that the receiver be turned off during transmission. Both operations are computer controlled. The receiver must be designed with as small an amount of electronic noise as possible, so that the radio frequency coil noise, an intrinsic restriction, rather than receiver noise, a technical restriction, remains the fundamental limit on noise.

COMPUTER SYSTEMS FOR NMR IMAGING

Although some requirements are common, different imaging techniques affect data acquisition and storage requirements differently. Whole-plane and whole-object techniques, in which all the data are collected at once, place more demands on the receiver than do single-line or single-point techniques. Computational requirements tend to increase accordingly, because, in general, the more data that are collected simultaneously, the more calculations that are needed to unfold them. However, current microprocessor technology makes it possible to operate in parallel several data-acquisition and data-processing units and, in addition, to expand storage capabilities. Therefore, we may expect that computer systems will not be a performance-limiting factor.

Because of the heavy involvement of x-ray CT manufacturers in the NMR field, NMR imagers have data systems derived from, and thus quite similar in architecture and appearance to, the data systems used in x-ray CT. One interesting aspect of NMR may be the advantageous use of color displays—the consensus being that color has not proved useful so far in either x-ray or nuclear medicine imaging.

The use of the color spectrum as an intensity scale is often more confusing than informative because we do not perceive a change in hue as a simple change in intensity. The transition between blue and green, for example, is perceived as more decisive than the transition between two shades of grey. Consequently, in a color image, a distinct edge is perceived whenever a small change in intensity involves a shift in the color spectrum. Therefore, to create meaningful color images, one must have an imaging system in which object contrast is great enough, and spatial resolution is fine enough, for shifts in color to give a fairly reliable indication of morphologic features. In CT, contrast between soft tissues is low, and color has been used to try to enhance this contrast. However, the application of the entire color spectrum to an objectively small range of intensities is misleading. Fractional shifts in the intensity scale are perceived as substantial shifts; edges are perceived where no morphologic distinctions exist. In nuclear medicine imaging, object contrast is large, but spatial resolution is poor. In a black and white image, poor spatial resolution will be perceived as blurring, a slow decline in the intensity scale from peak value to background value. In the color image, however, the decline in intensity is indicated by transitions through many colors, each of which the eye perceives as distinct. Thus, what in the black-and-white image would be perceived as inadequate delineation, in the color image is perceived as excessive, and falsely significant, delineation. NMR combines the high contrast of nuclear medicine with spatial resolution comparable to that of x-ray CT; therefore, color is not likely to produce false edges or confusing features and could well become a useful technique to highlight the specificity of NMR images.

Another area of potential use for color displays is to synthesize the different parameters provided by NMR images (hydrogen density and T_1 and T_2 relaxation times). If the composite image is in black and white, features that are appreciably enhanced by use of one or the other of the imaging parameters can be tinted to combine in one image the information provided from several images. Color can be used in a similar fashion to synthesize information from NMR images of other endogenous elements with the hydrogen image. Because of the variety of information that NMR images can provide, it is important to develop color display methods that allow for the concise and meaningful transmission of this information.

FINAL CONSIDERATIONS

One of the distinctive features of NMR imaging is that if the most expensive components of the system (main magnet, gradient coil, power supplies, and computer system) are built with adequate flexibility and capabilities, they need not be changed as successive "generations" of imaging methodology are cycled because the "software" nature of the data-generation and data-acquisition process will allow for improvements to be made through relatively minor changes in hardware, thereby extending the useful lifetime of NMR systems. This increased lifetime will be aided by the lack of moving parts (except for the bed). As a consequence, in spite of initial costs comparable to those of advanced x-ray CT systems, NMR imagers might provide lower life-cycle costs.

BIBLIOGRAPHY

Becker JJ: Permanent magnets. *Sci Am* 223:92–100, 1970.

Bottomley PA, Andrew ER: RF magnetic field penetration, phase shift, and power dissipation in biological tissue: Implications for NMR imaging. *Phys Med Biol* 23:333–347, 1978.

Garrett MW: Axially symmetric systems for generating and measuring magnetic fields. *J Appl Phys* 22:1091–1107, 1951.

Geballe TH, Hulm JK: Superconductors in electric power technology. *Sci Am* 243:138–172, 1980.

Hoult DI, Lauterbur PC: The sensitivity of the zeugmatographic experiment involving human samples. *J Magn Reson* 34:425–433, 1979.

Hulm JK, Kunzler JE, Matthias BT: The road to superconducting materials. *Phys Today* :34–43, January 1981.

Hulm JK, Matthias BT: High field, high-current superconductors. *Science* 208:881–887, 1980.

Kolm HH, Freeman AJ: Intense magnetic fields. *Sci Am* 212:66–78, 1965.

Kunzler JE, Tanenbaum M: Superconducting magnets. *Sci Am* 206:60–67, 1952.

Montgomery DB: *Solenoid Magnet Design: The Magnetic and Mechanical Aspects of Resistive and Superconducting Systems.* New York, Wiley-Interscience, 1969.

Terman FE: *Radio Engineer's Handbook*, ed 1. New York, McGraw-Hill, 1943, pp 73–90.

II

Imaging

5

NMR Characteristics of Normal and Abnormal Rat Tissues

Peter L. Davis, M.D.

Leon Kaufman, Ph.D.

Lawrence E. Crooks, Ph.D.

Alexander R. Margulis, M.D.

The content of NMR images will be substantially different from that of diagnostic images already familiar to radiologists. Whereas conventional x-ray radiographic and computed tomogram images depend on electron density, a physical parameter that is reasonably well understood, NMR images depend on the density of hydrogen, its state of motion, and its magnetic relaxation times, T_1 and T_2 (1, Chapters 2 and 3). The amount of hydrogen present and its state of motion are parameters that are relatively easy to measure in experimental models; however, the magnetic relaxation times of the hydrogen in different tissues are not easily predictable, and their diagnostic implications are poorly understood. Nevertheless, it is the effect of these relaxation times on the observed NMR image that may allow these images to have high soft-tissue specificity, and it is precisely this possibility that has attracted the radiologic community's attention to the development of NMR imaging systems.

Due to the dependence of NMR images on the chosen NMR imaging technique, the tissue specificity of NMR images must be determined as part of the general introduction of NMR instrumentation into diagnostic practice. In addition, if NMR images are ever to be interpreted on other than purely empiric grounds, basic research must be done to determine the physical and physiologic causes of the information NMR images provide. The studies presented here are part of a continuing series being done at the University of California at San Francisco to help determine the meaning and extent of the information provided by NMR images.

NMR INTENSITY IMAGES OF NORMAL RAT ANATOMY

Instrumentation

The NMR imager has been described elsewhere (2). It has a magnetic field of 0.35 Tesla, which corresponds to a hydrogen nucleus resonant frequency of 15 MHz. Data are collected line by line. The line, which is oriented in the transverse direction, is moved electronically from the top to the bottom of the animal. The sections are 4.2 or 8.4 mm thick, and spatial resolution is 2×1 mm in the imaging plane. The aperture of the imager is 6.5 cm.

Any one image is a two-dimensional representation of the NMR intensity (I) under the given imaging conditions. In the imager used here, I is given by the expression

$$I = H\mathrm{f}(v)\,[\exp(-a/T_2)]\,[1-\exp(-b/T_1)], \qquad [1]$$

where I is the NMR intensity in a particular region of the image, H is the local hydrogen density, a is the T_2 parameter of the instrument, measured in milliseconds and varied within a broad range under computer control, b is the T_1 parameter of the instrument, measured in seconds and also computer controlled, and $\mathrm{f}(v)$ is a function of both the speed with which hydrogen nuclei move through the region being imaged and of the fraction of the total number of nuclei that are moving.

Normal Rat Tissues

To determine first the extent of the anatomic information NMR images provide, images of normal rats have been correlated with corresponding frozen cross sections of the animals (3). The images presented here (Figs. 5.1-5.4) illustrate the appearance of organs in the normal rat. All were

---

Figure 5.1. Cross sections through the head of a live rat. (a) NMR tomogram of the rat. The most intense organ is the brain. It is surrounded by a dark outline, which is the skull. The mandible is also dark. The muscles of the jaw and tongue are intermediate in intensity, whereas the air in the nasopharynx provides the least intense signal. (b) Computed tomogram of the live rat. (c) Anatomic section. (d) Radiograph of the anatomic section.

Figure 5.2. Cross sections through the upper abdomen of a live rat. (a) NMR tomogram of the rat. The subcutaneous fat is most intense. The liver is next highest in intensity. The spinal cord is not as intense as the brain, probably due to partial volume averaging with the adjacent vertebrae. (b) Computed tomogram of the live rat. (c) Anatomic section. (d) Radiograph of the anatomic section.

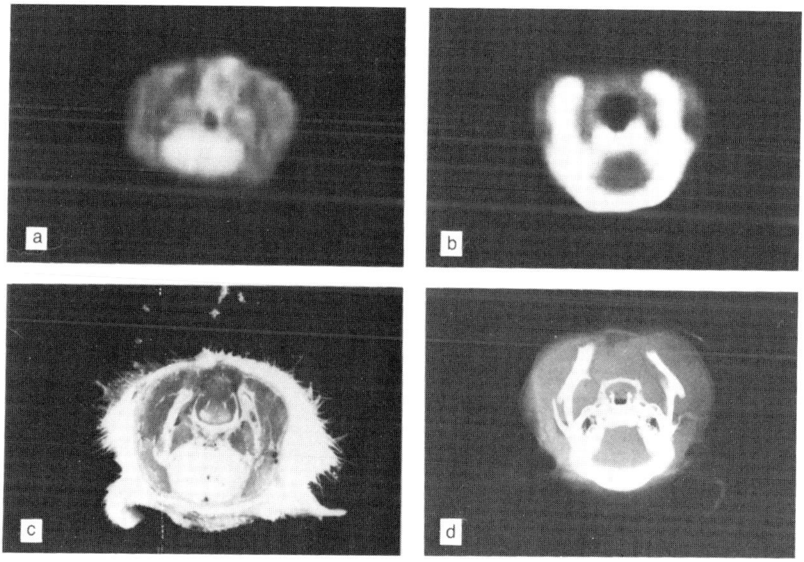

Figure 5.1. (legend on facing page)

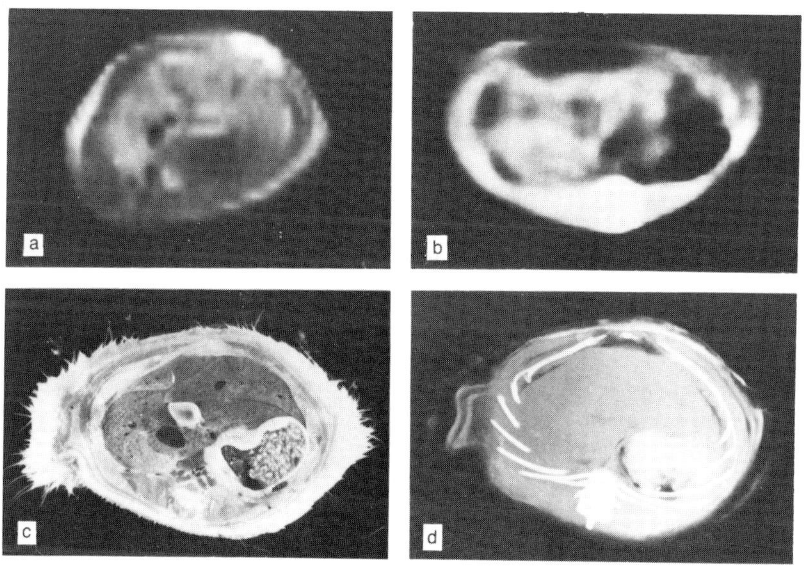

Figure 5.2. (legend on facing page)

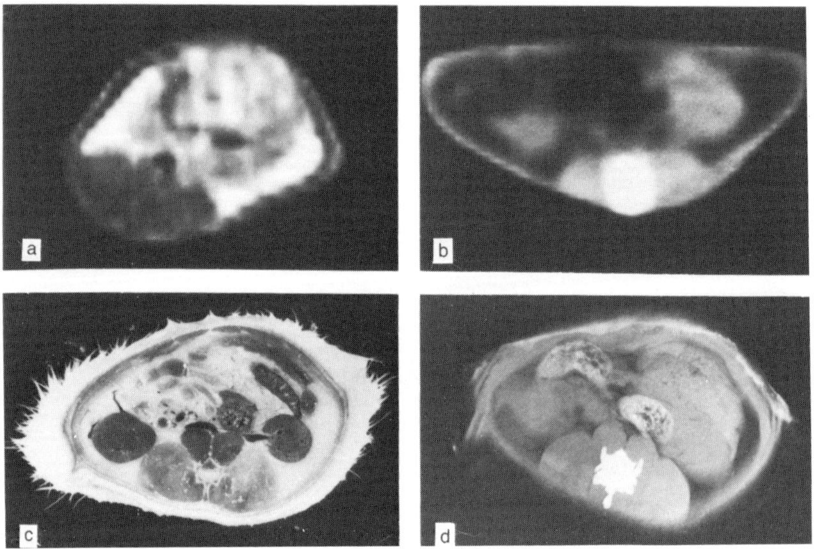

Figure 5.3. Cross section through the mid-abdomen of a live rat. (a) NMR tomogram of the rat. The kidneys are gray, surrounded by an intense area of perinephric fat. The air and fluid in the bowel are dark, as is the flowing blood in the aorta. (b) Computed tomogram of the live rat. (c) Anatomic section. (d) Radiograph of the anatomic section.

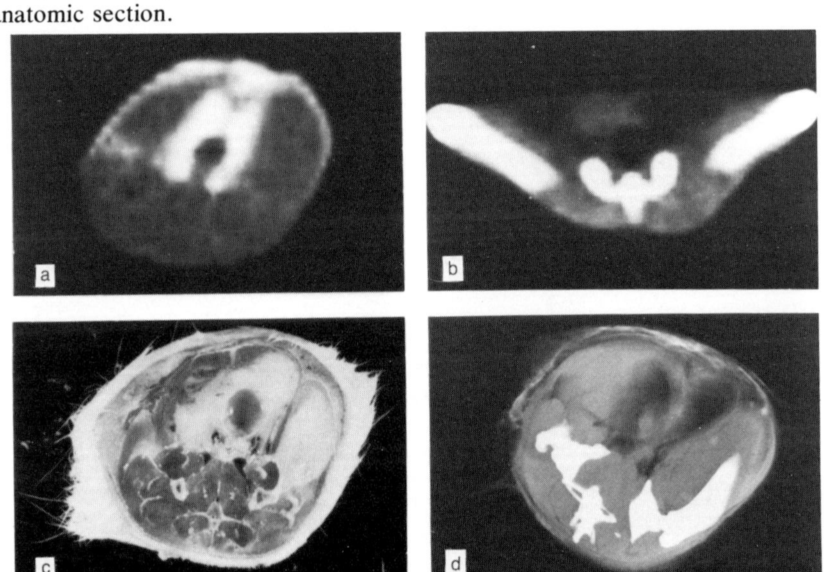

Figure 5.4. Cross section through the pelvis of a live rat. (a) NMR tomogram of the rat. The urinary bladder is surrounded by fat. The tips of the iliac crest are seen as round, dark gray areas within the flank muscles. (b) Computed tomogram of the live rat. (c) Anatomic section. (d) Radiograph of the anatomic section.

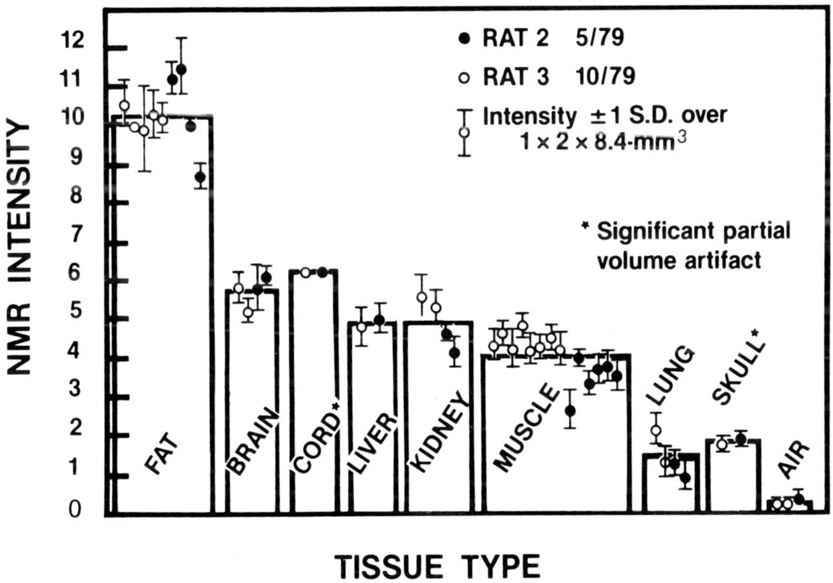

Figure 5.5. Bar graph of NMR intensities for various organs of two live rats, imaged in a 5-month interval. Good agreement is shown between the two animals. Error bars represent 1 SD for a 2 × 1-mm area.

taken with the instrument's T_1 parameter, b, set at 0.5 sec and the instrument's T_2 parameter, a, set at approximately 28 msec. These settings produce a gray scale in which the relative intensities of tissues are, in order of decreasing intensity, fat, tissues containing lipids, such as brain and spinal cord, nonfatty tissues, lung (which is partially air), bone (which is not completely dark because of partial volume effects), and air (Fig. 5.5). Tissue contrast is high: For example, contrast between fat and muscle is 250%.

INTENSITY AND CALCULATED NMR IMAGES

Intensity Images: T_1 and T_2 Dependencies

The gray scale just illustrated by the preceding images depends on several variables, of which the most important are the instrument T_1 and T_2

parameters. Referring back to Equation 1, it is apparent that when the T_2 of a tissue is long and the T_1 is short, its intensity is high (as in fat); low intensities occur when the converse is true. Changing the T_1 or T_2 instrument parameters will discriminate between tissues according to the relative lengths of their relaxation times (Fig. 5.6).

For example, muscle, which has a short T_2 time and a long T_1 time, will become brighter as the instrument's T_1 parameter is lengthened but darker when the instrument's T_2 parameter is lengthened. As well as providing an absolute change in NMR intensities, varying the T_1 and T_2 parameters can also selectively enhance the contrast between tissues. For example, if the T_2 parameter is lengthened, although both fat and muscle will decrease in absolute intensities, the contrast between them will increase. Consequently, signal to noise (which depends on contrast and intensity) can be made maximal by a judicious selection of a and b parameters.

After gaining experience with the imager, we now produce four images with $a = 28$ and 56 msec and $b = 0.5$ and 1.0 sec. During one imaging sequence, we generate two images simultaneously, each for a different value of the a parameter. To obtain images with a different value of b, the imaging sequence is repeated.

Obtaining NMR images at several instrument T_1 and T_2 parameter settings is essential because at certain settings, pathologic lesions may blend into the surrounding tissues and be missed (Fig. 5.7).

Calculated Images: T_1, T_2, H, and A

Using these four standard images of NMR intensity, we calculate images of the T_1 and T_2 relaxation times and $H\mathrm{f}(v)$ using the relationship described in Equation 1. We also calculate a fourth image, the asymptote (A). This is an intensity image with the parameter b set at $b = \infty$, that is, an

---

Figure 5.6. Four images through the abdomen of a live rat, obtained with the T_1 and T_2 parameters indicated. Note the enhancement of fatty tissues when the T_2 parameter is increased. The data in these four images can be used to obtain T_1, T_2, and hydrogen images.

Figure 5.7. Images through the flanks of a live, supine rat with a sarcoma implanted in one flank. T_1 values increase from left to right, and from top to bottom, from 0.25 to 1.5 sec.

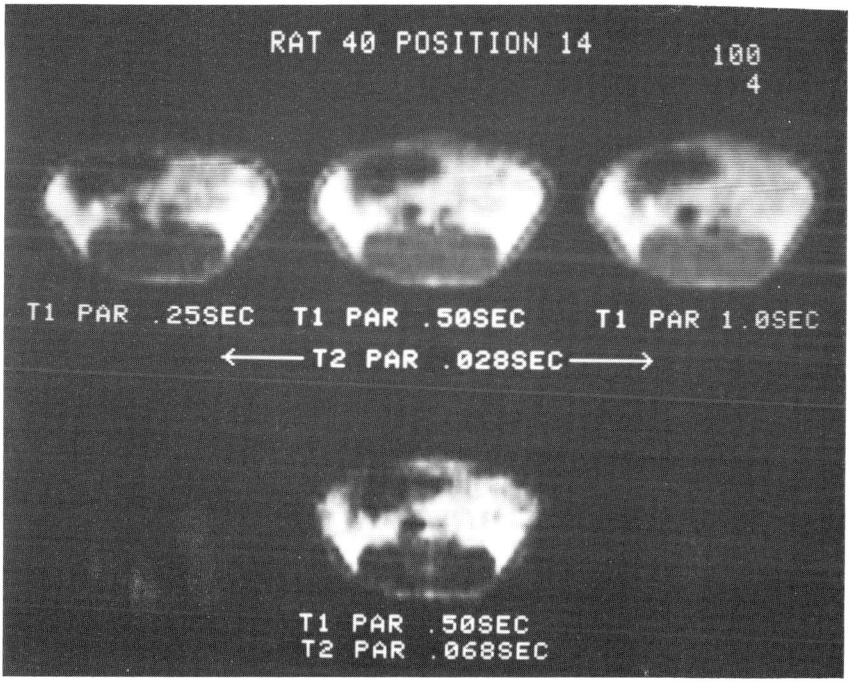

Figure 5.6. (legend on facing page)

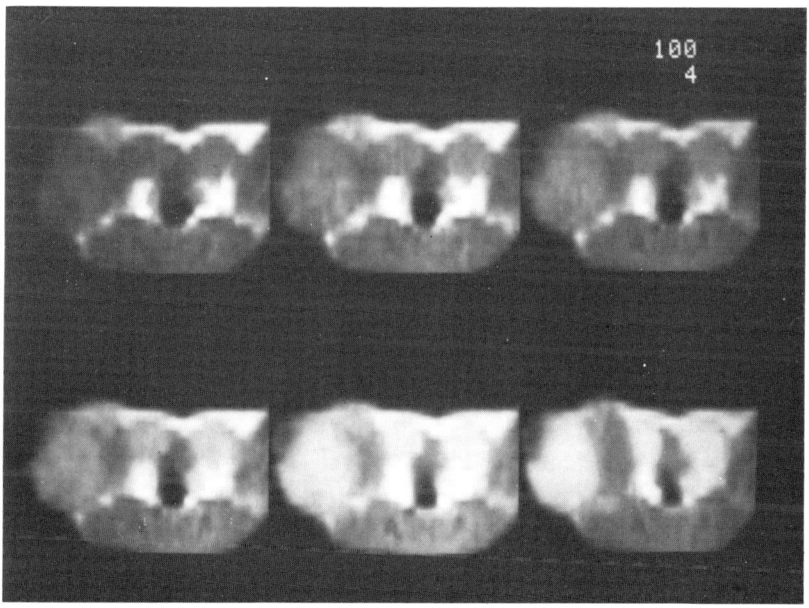

Figure 5.7. (legend on facing page)

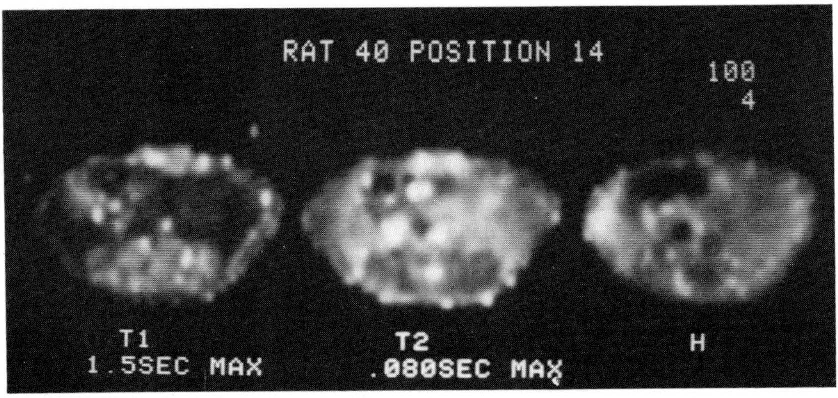

Figure 5.8. T_1, T_2, and hydrogen (H) images obtained from the four images shown in Figure 5.6. The T_1 and T_2 images are shown with gray scales ranging from 0 to 1.5 sec and from 0 to 80 msec, respectively. The H image is in arbitrary intensity units, not yet calibrated for hydrogen density. These images have been smoothed with a 5-point operator. Note the short T_1 and long T_2 values of fat and the long T_1 value of muscle. Spinal cord also has a long T_2 value, and the spine is observed as a void in the H image.

image with T_2 and $H\text{f}(v)$ but no T_1 dependence. These images are calculated from the intensity images pixel by pixel (Fig. 5.8).

Three major sources of systematic errors enter into these calculations: partial-volume effects, which for an 8.4 mm thick slice of the rat can sometimes be important; the representation of tissue relaxation times by single exponentials, which, in fact, are multiple-component functions; and flow of blood through the tissues, which brings into the resolved element hydrogen that has not undergone the same irradiation sequence as stationary hydrogen. As a consequence of these errors, values obtained from Equation 1, although specific to the imager, are not necessarily reproducible in other systems.

Even so, because of the exponential nature of the decay times, the qualitative relationship should be comparable (e.g., fat will always have a shorter T_1 and a longer T_2 relaxation time than will other normal tissues). Once the T_1 and T_2 dependence of any imaging technique has been determined, H, T_1, and T_2 images can be calculated, although, as has already been discussed in Chapter 2, the ease with which the images can be calculated varies widely.

NMR IMAGES OF DISEASE

Potential Diagnostic Specificity of NMR

Due to previous NMR in vitro research, the possibility that NMR imaging may differentiate malignant from benign processes better than other noninvasive diagnostic modalities is already of particular interest. In vitro measurements in rats (4), and later in human beings (5-7), have indicated differences in the T_1 and T_2 relaxation times between normal and cancerous tissues. The exact interpretation of these observations is unclear; however, it seems to point toward a correlation with degree of hydration of tissue (8,9). These results have not always been consistent (7). It has been pointed out that controls have been few or absent (6) and that edema would produce the same effect as malignant processes (10). The suggestion that there are substantial differences between cancerous and normal tissues has been disputed, with some data indicating a distinct overlap of relaxation times (11,12). In studies that process tissues for in vitro analysis, investigators conclude that "as far as differentiating between malignant tissue, great care must be taken in discriminating between effects caused by temperature or necrosis and those resulting from true changes related to malignant growth" (12).

Obviously, an important control process when studying the diagnostic capabilities of NMR images, one that has not always been used in in vitro studies, is not only to compare normal and abnormal tissues but also to compare the NMR characteristics of different abnormalities to see if benign and malignant processes can be distinguished. Increasing our understanding of the physiologic basis of NMR characteristics of tissues should also increase the diagnostic specificity of the technique. Therefore, in the first stage of our work with NMR characterization of pathologic lesions, a broad spectrum of disease, including benign and malignant tumors, abscesses, and hematomas, was induced in rats and imaged (13). Lesions were chosen as being representative of the types of lesions NMR imaging would be used to detect. Correlations were made between normal and abnormal tissues and their magnetic relaxation times, then between the relaxation times of the normal and abnormal tissues and their water, extracellular fluid, and plasma contents. In the second study, a single tumor strain was used to study the variability in NMR responses found in a homogeneous tumor model (14).

Normal and Abnormal Tissues: T_1 and T_2 Correlations

Figures 5.9, 5.10, and 5.11 show the characteristics of several tumors. Figures 5.12 and 5.13 show the findings in two abscesses (one pyogenic and the other sterile). Figure 5.14 shows a subcutaneous hematoma. Reviewing these images, one sees that the lesions can be identified on some of the NMR intensity images and some of the calculated images. Specifically, all the lesions can be discriminated from surrounding tissues in either the T_1 or T_2 time image; some lesions can be discriminated in both. Sometimes the lesion may appear similar to the surrounding area in one or the other of the two relaxation time images, but taking both together, all lesions can be clearly distinguished from the surrounding normal tissues, although, as we show below, the T_1 and T_2 values of hematomas and abscesses can overlap the T_1 and T_2 values of tumors, thus precluding unique characterization of diseases.

Plotting the values of T_1 and T_2 for each tissue (Fig. 5.15), we find that, except for tumors, each tissue type occupies a relatively distinct region. This observation is better illustrated in Figure 5.16, which shows the mean and standard deviations of the T_1 and T_2 times for each tissue. Each rectangle contains 90% of all values for that tissue type.

Tumor values, however, cover a wide range, which can overlap that of both normal and other abnormal tissues. This overlap should be expected because the composition of tumors can vary widely. Even when imaging a single tumor model (no. 394A hepatoma) in one rat line (ACI/N), great variability in T_1 and T_2 is observed among tumors (Fig. 5.17). In this study (14), which was correlated with results of pathologic examination, all tumors larger than 8 mm^2 in area were detected. T_1 and T_2 did not correlate with size (Fig. 5.18), indicating that for the detected tumors, partial-volume artifacts were not important. Even for this relatively uniform tumor model, 30% of the T_1 and T_2 values overlap areas that contain 90% of the values for abscesses, hematomas, and normal brain, as determined in the previous study.

Figure 5.9a. NMR images through the upper pelvis of a rat with a calcified fibroma. From left to right, $b = 0.25$, 0.5, and 1.0 sec, respectively. Top and bottom rows are for $a = 28$ and 56 msec, respectively. The tumor is seen as a low-intensity region.

Figure 5.9b. Calculated images of T_1, T_2, hydrogen (H), and asymptote (A) showing the tumor as a low-intensity region in all four images. No other tissue shows the same characteristics in all four images.

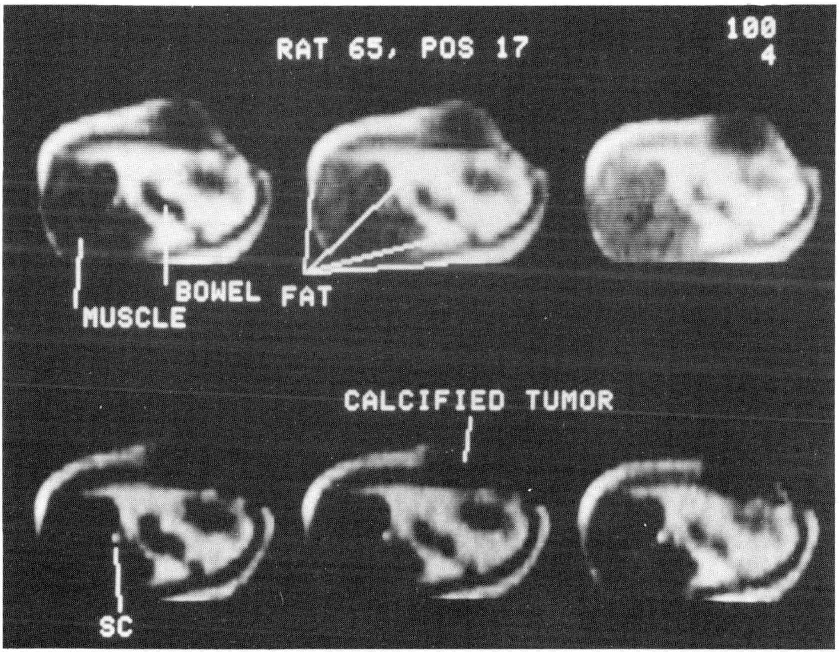

Figure 5.9a. (legend on facing page)

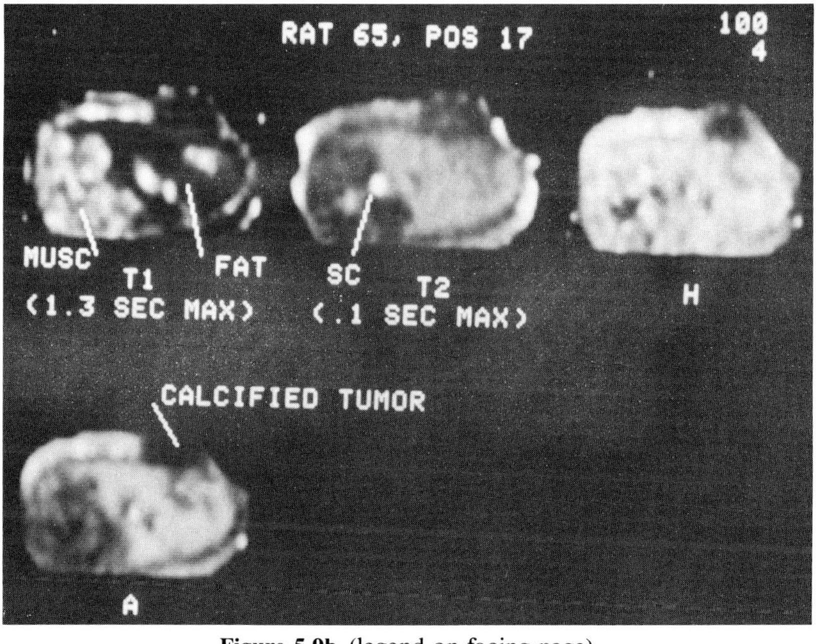

Figure 5.9b. (legend on facing page)

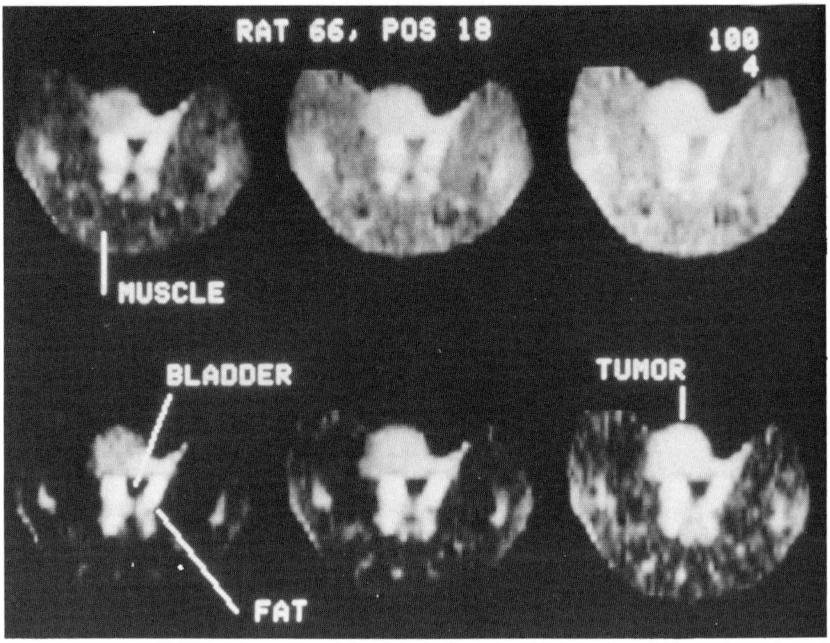

Figure 5.10a. NMR images (see Fig. 5.9a) through the lower pelvis and flanks of a rat with an adenocarcinoma located ventral to the bladder.

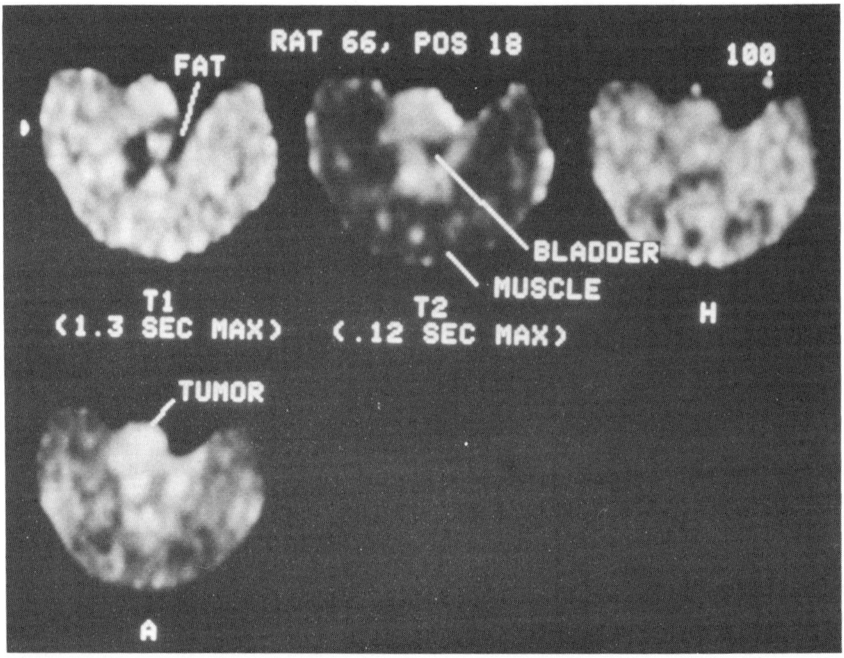

Figure 5.10b. Calculated images showing tumor with long T_1 and T_2 values. Fat has similar T_2 but shorter T_1, whereas muscle shows the converse.

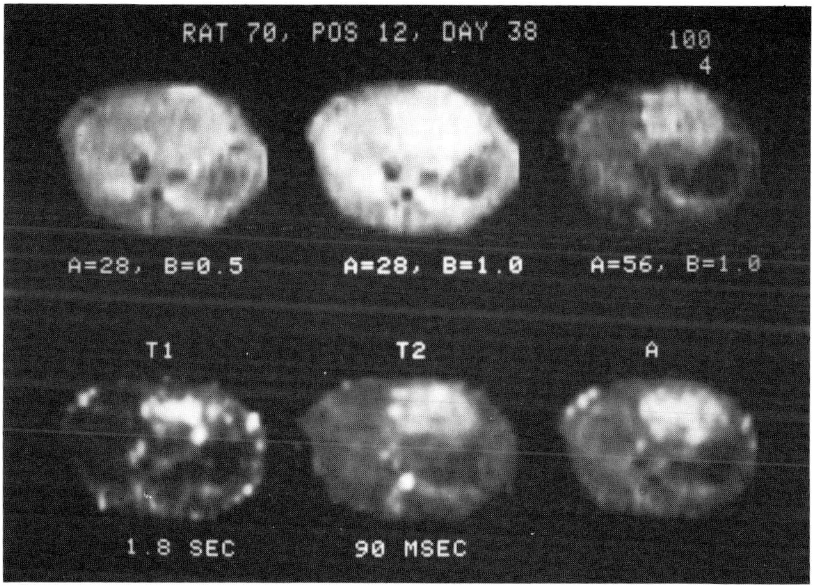

Figure 5.11a. NMR images through the liver of a rat show tumor in the left half of the liver. Both T_1 and asymptote images demonstrate a more intense core in the tumor.

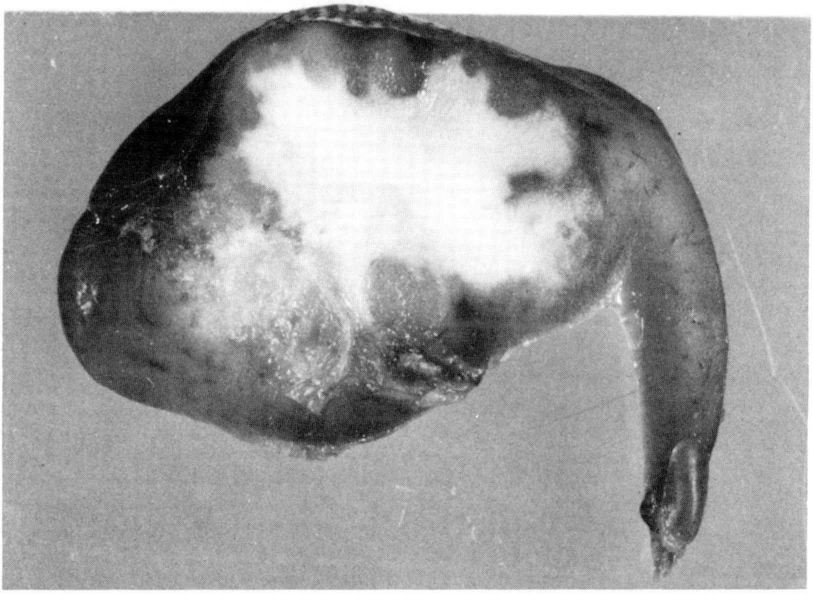

Figure 5.11b. The intense core in the tumor is also seen in the pathology section as a white core surrounded by a slightly darker region.

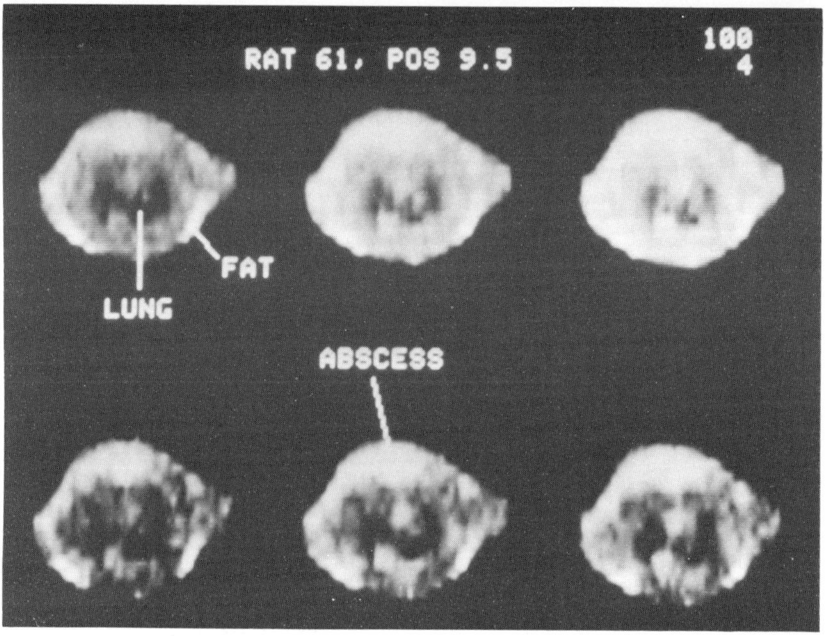

Figure 5.12a. Intensity images through the apices of the lungs of a rat with a pyogenic abscess in the nape of the neck.

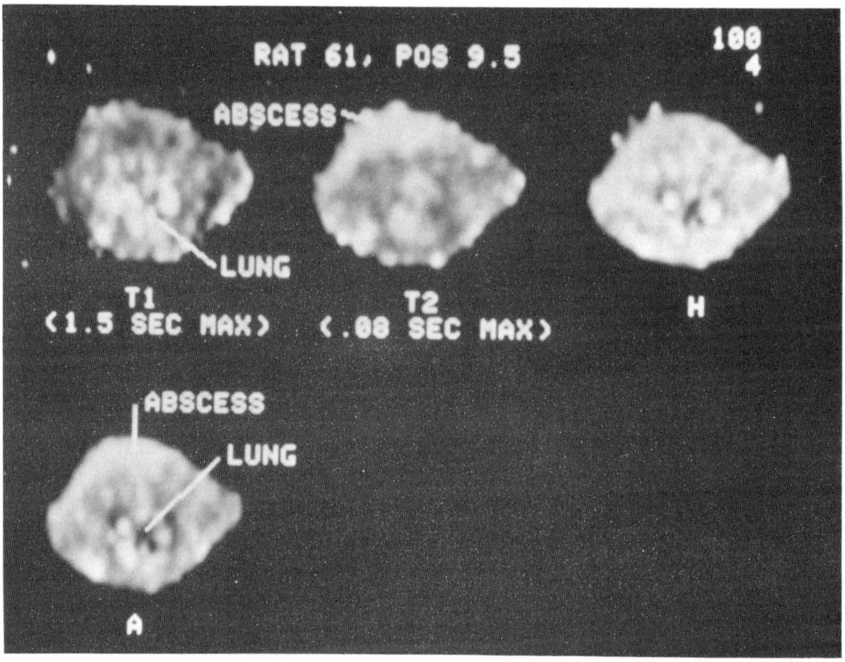

Figure 5.12b. Calculated images of the same abscess. Note short T_1 and long T_2 times of the abscess.

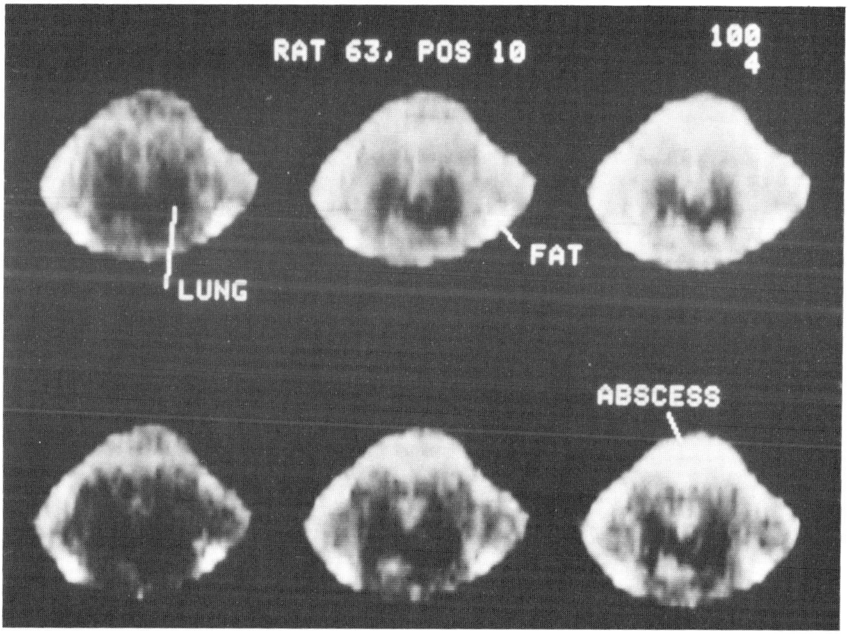

Figure 5.13a. Intensity images through the apices of the lung of a rat with a sterile abscess in the nape of the neck.

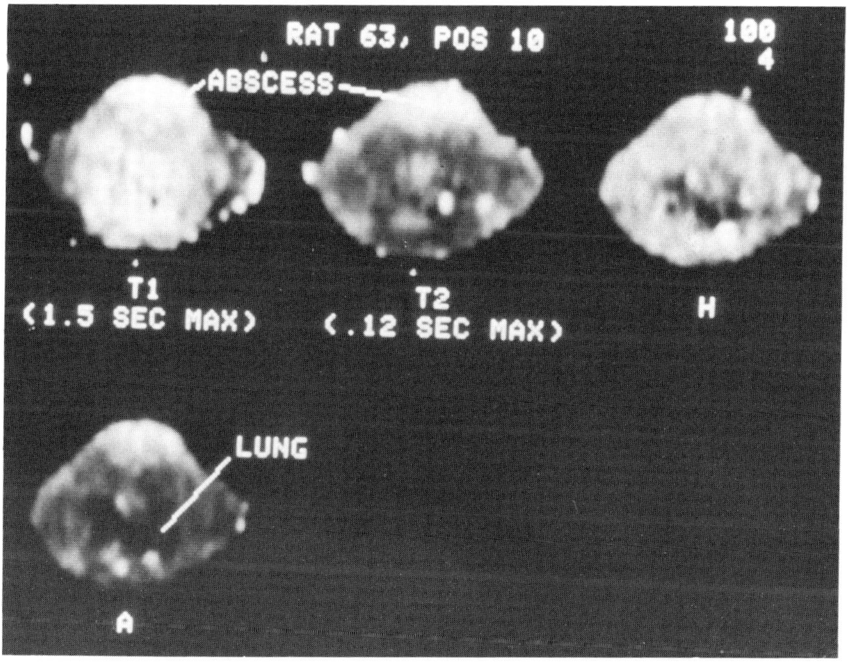

Figure 5.13b. Calculated images of the same abscess. In this case, both T_1 and T_2 times of the abscess are long.

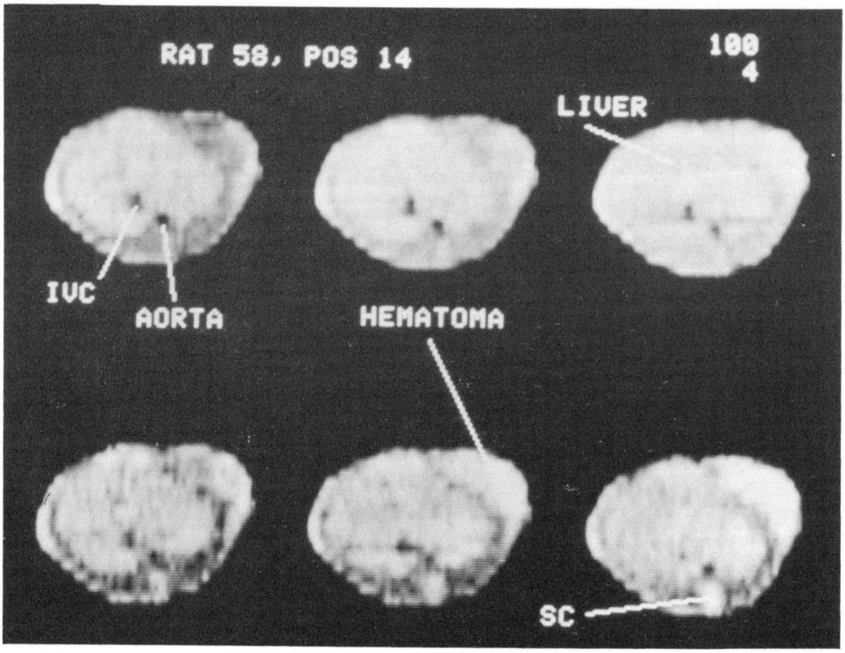

Figure 5.14a. (Legend on facing page)

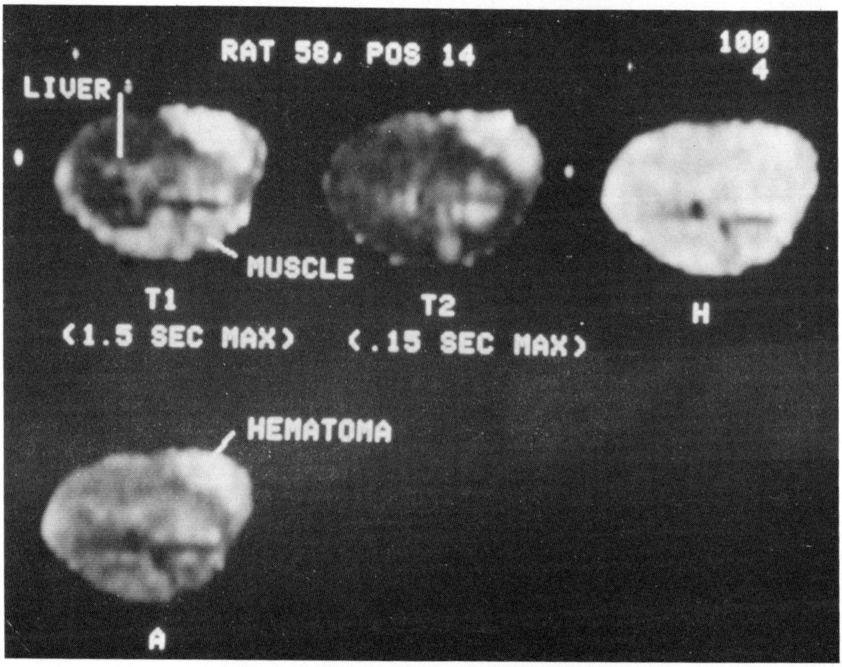

Figure 5.14b. (Legend on facing page)

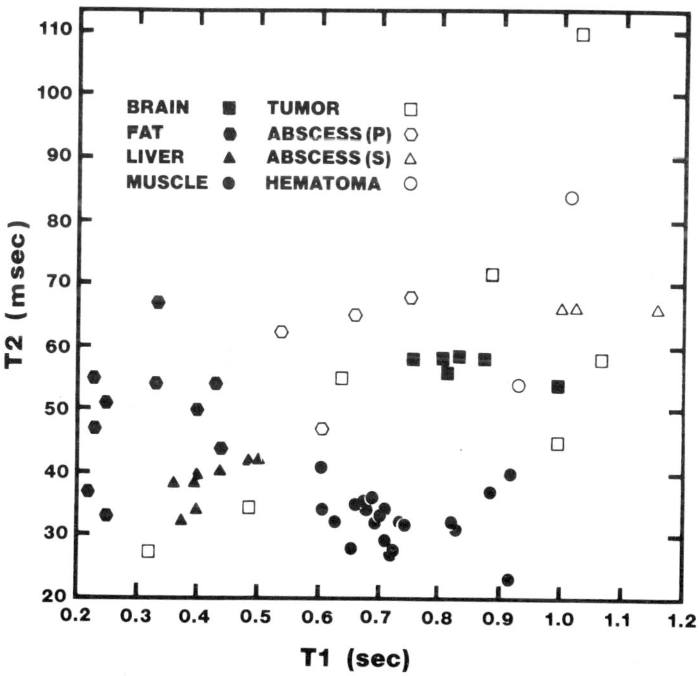

Figure 5.15. Comparison of T_2 and T_1 times for various normal and abnormal tissues in live rats. P, pyogenic abscess; S, sterile abscess.

Figure 5.14a. Intensity images through the upper abdomen of a rat with a subcutaneous hematoma.

Figure 5.14b. Calculated images of the subcutaneous hematoma show that the hematoma has long T_1 and T_2 times.

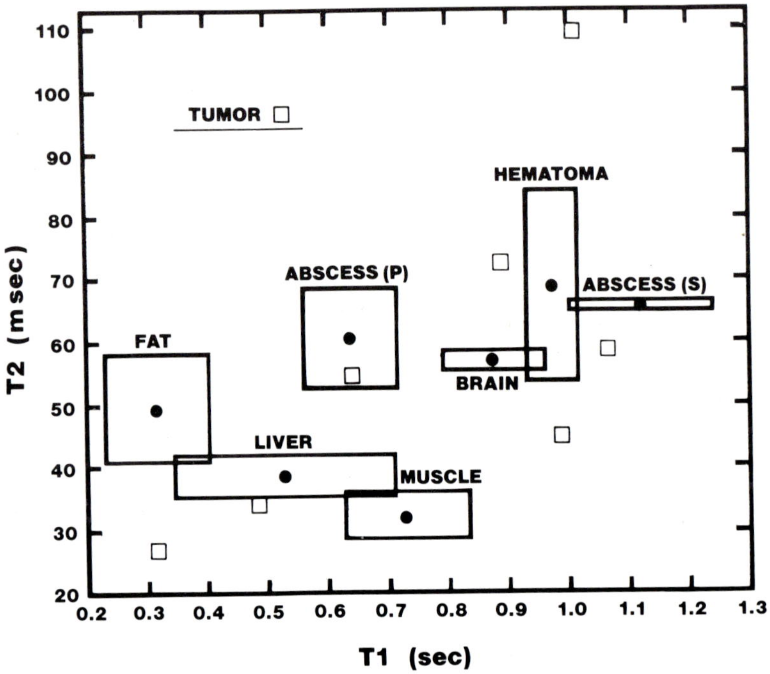

Figure 5.16. Mean and standard deviations of T_1 and T_2 of various tissues in a live rat. Individual values of tumors are indicated by small squares.

The ability to differentiate lesions from surrounding normal tissue is, obviously, of decisive importance to any potential diagnostic technique. When plotting the contrast of T_1 and T_2 of abnormal tissue against the T_1 and T_2 of normal tissue, where contrast is calculated as

$$[(\text{Relaxation time of abnormal tissue/Relaxation time of normal surrounding tissue}) - 1], \quad [2]$$

no points fell at the zero intersection, where no contrast exists between abnormal and normal tissues (Figs. 5.19a and 5.19b). In addition, the plot also demonstrates that most of the pathologic tissues will appear more intense than the surrounding tissues on at least one of the two images. The only exceptions to this rule were the two points in the lower left, which were darker than those of surrounding tissues on both relaxation time images. These points represented fibrotic tumors (Fig. 5.9).

Physiologic Correlations

NMR parameters obtained from images of live animals were correlated with water content, extracellular fluid volume, and plasma concentrations. Plots were made of the tissue rate constants $1/T_1$ and $1/T_2$ versus the physiologic parameters.

Both rate constants vary in direct proportion to water content (Figs. 5.20 and 5.21), with the tumors being part of the continuum. The rate constants, $1/T_1$ and $1/T_2$, correlate with water, with r values of 0.81 and 0.79, respectively. The only anomalous behavior is that of the T_2 of fat, which has a T_2 value longer than that expected on the basis of water content alone.

Figures 5.22 and 5.23 show the relationship between rate constants and extracellular fluid. Behavior similar to that seen in Figures 5.20 and 5.21 is found, indicating that it is the water content itself, rather than its distribution, that affects relaxation times. Scatter of data points is larger than for water. This finding is not unexpected because extracellular fluid is a smaller subset of the total water content. Plasma content does not correlate with NMR parameters well, probably because it occupies only a small fraction of the total tissue volume.

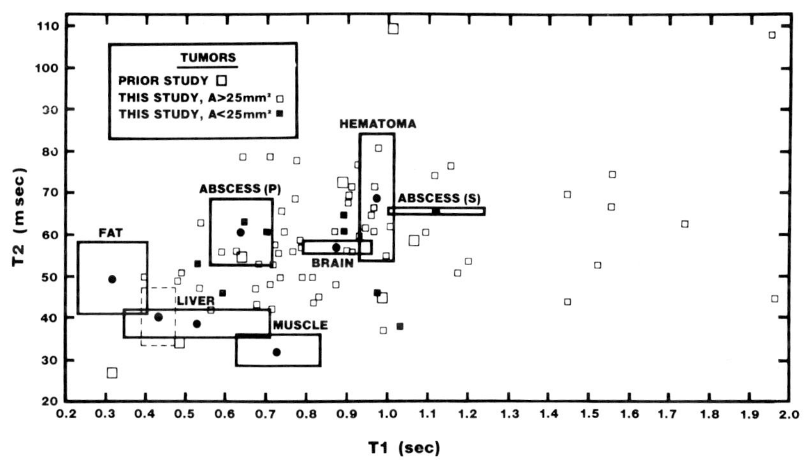

Figure 5.17. Comparison of T_1 and T_2 relaxation times for various tissues. Large rectangles enclose 90% of values for each kind of tissue. Dashed rectangle shows normal liver values for hepatoma study. The tumors for Figure 5.16 are also shown here. The smallest squares are for the hepatoma study. Solid squares are for hepatomas of less than 25 mm^2 area in the image; open squares are for hepatomas between 25 and 370 mm^2 in area. Extrahepatic tumors (in the lung and skin) tended to have longer T_1 and T_2 values than those growing in the liver.

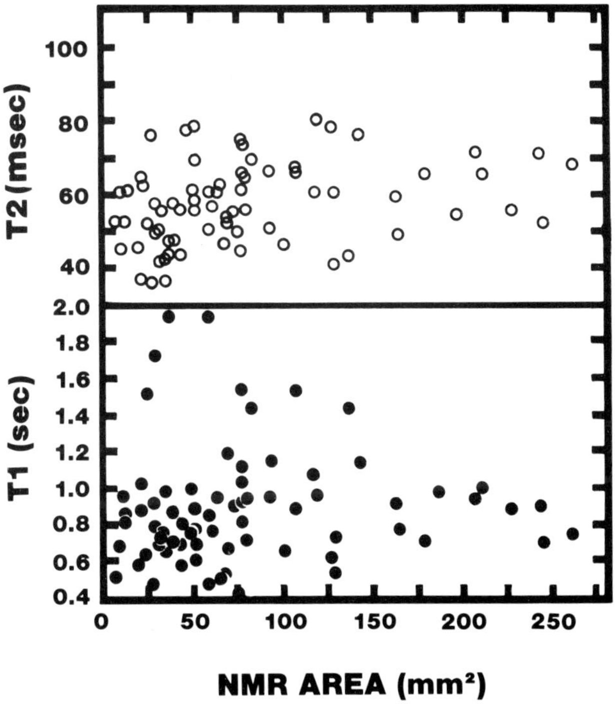

Figure 5.18. T_1 and T_2 values of tumors as a function of their NMR image areas.

Figure 5.19a. T_2 as compared with T_1 contrast for the various abnormal and normal tissues in live animals for the study in Figure 5.16. Abnormal tissues can be differentiated from normal, surrounding tissues only if both T_1 and T_2 contrast values are taken into account. The 0,0 intersection is the point where no contrast differentiation is observed.

Figure 5.19b. Contrast for hepatoma tumors shown in Figure 5.17.

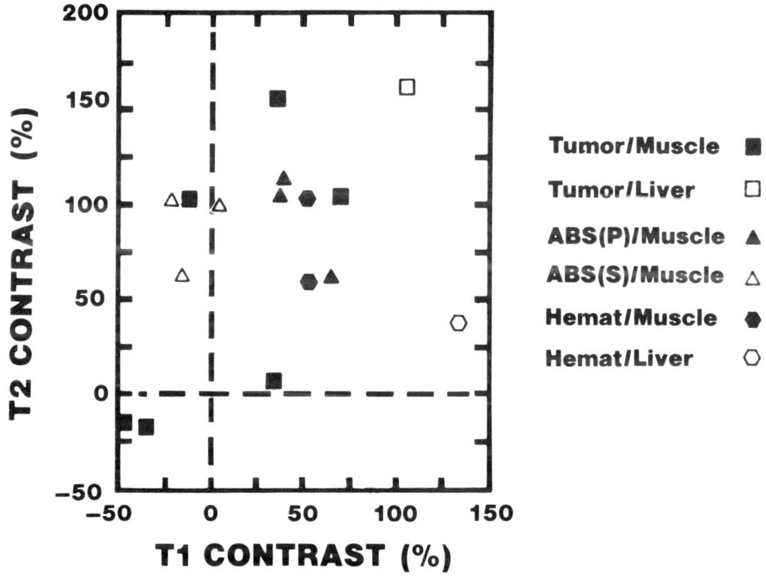

Figure 5.19a.(Legend on facing page)

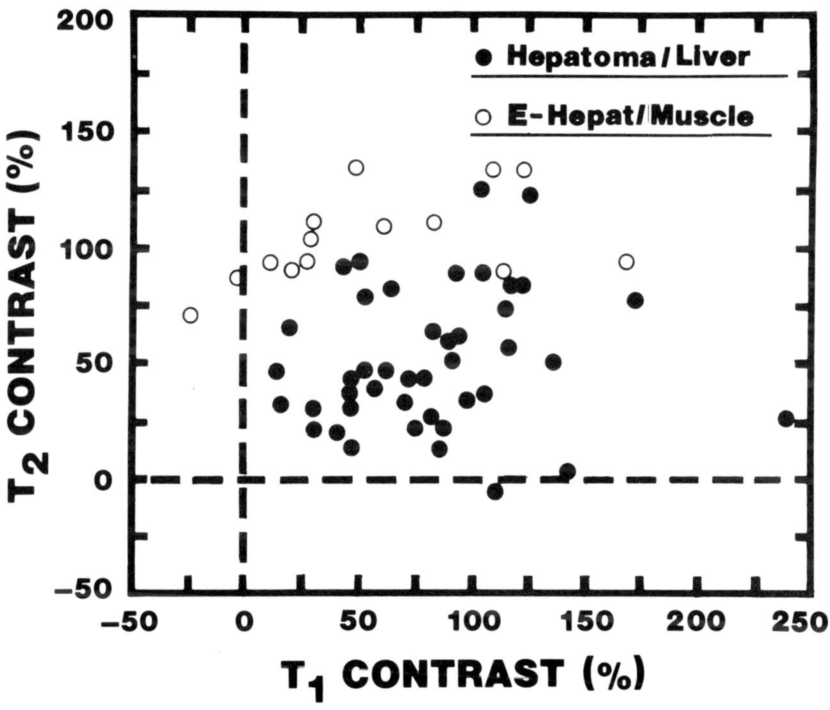

Figure 5.19b.(Legend on facing page)

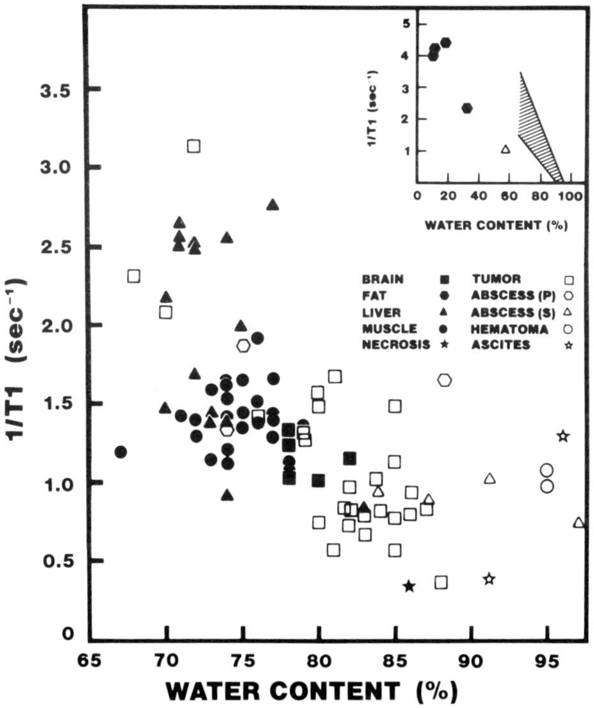

Figure 5.20. (Legend on facing page)

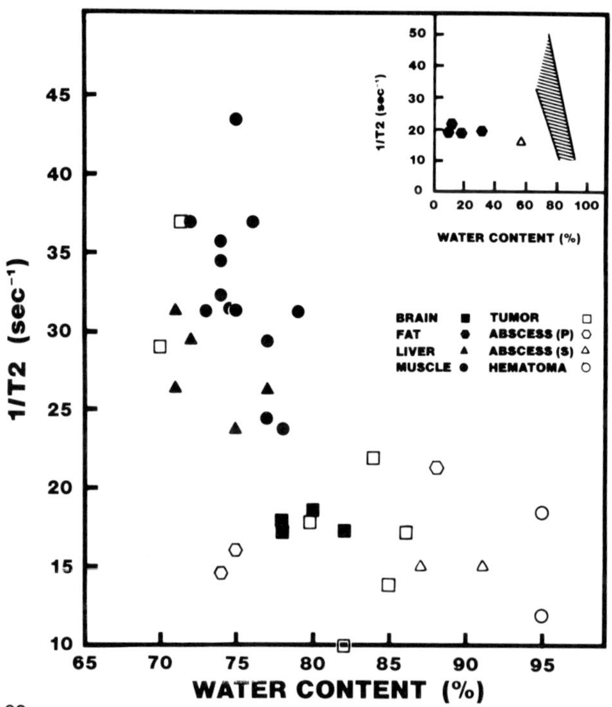

Figure 5.21. (Legend on facing page)

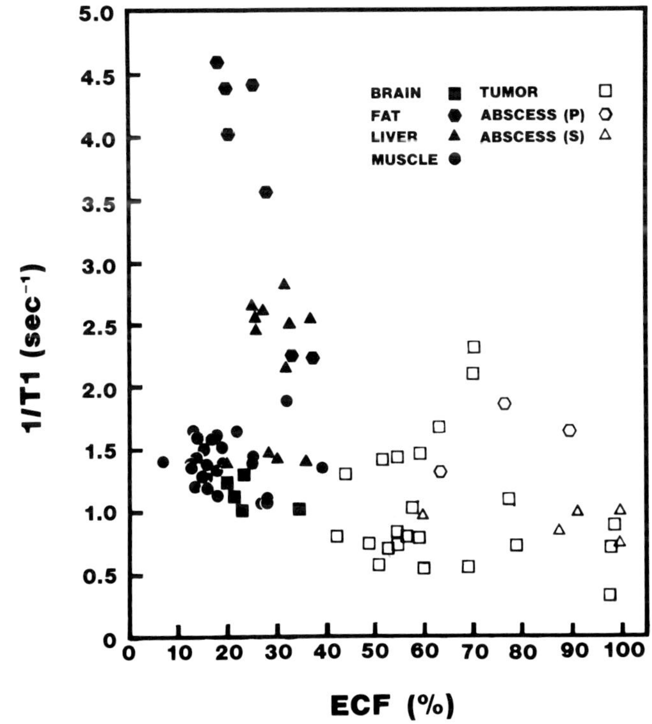

Figure 5.22. Rate constant $1/T_1$ versus extracellular fluid for various tissues.

Figure 5.20. Rate constant $1/T_1$ in a live animal as compared with water content shown for various normal and abnormal tissues. Inset shows the full range of water content, with the shaded region representing the main grouping of data. Fat follows the trend of other tissues.

Figure 5.21. Rate constant $1/T_2$ versus water content. Fat, shown in inset, does not follow the trend of other tissues.

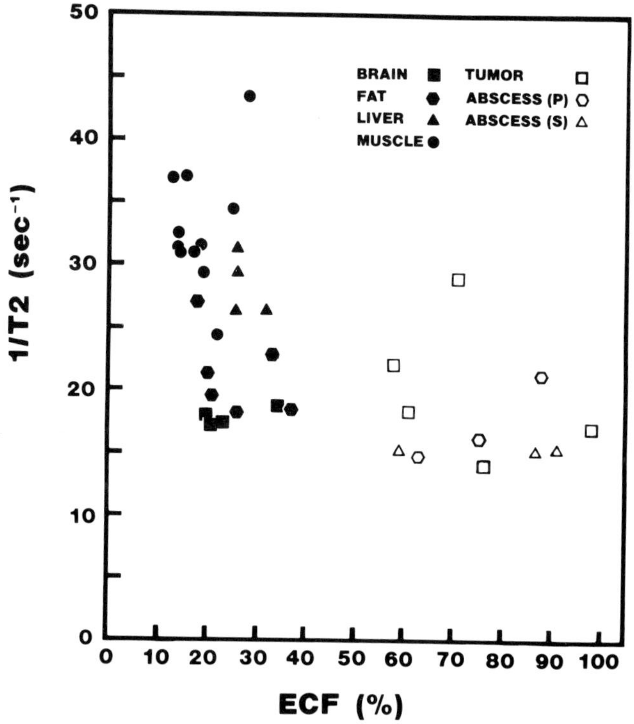

Figure 5.23. Rate constant $1/T_2$ versus extracellular fluid for various tissues.

BLOOD FLOW, RAT

One of the more promising attributes of NMR imaging is its sensitivity to blood flow, at least with some imaging techniques. Chapter 7 is devoted to a discussion of obtaining blood flow measurements by NMR. We present here some of the effects NMR's sensitivity to flow can have on the appearance of in vivo images.

Nuclei traveling through the volume elements being imaged experience imaging sequences that are different than those undergone by stationary nuclei. The fraction of nuclei leaving the selected volume in the time it takes to obtain information from it (approximately 35–70 msec in our case) will not yield any NMR signal; thus, with increasing flow speed, less nuclei remain within the imaging volume during the sequence time, and the signal drops.

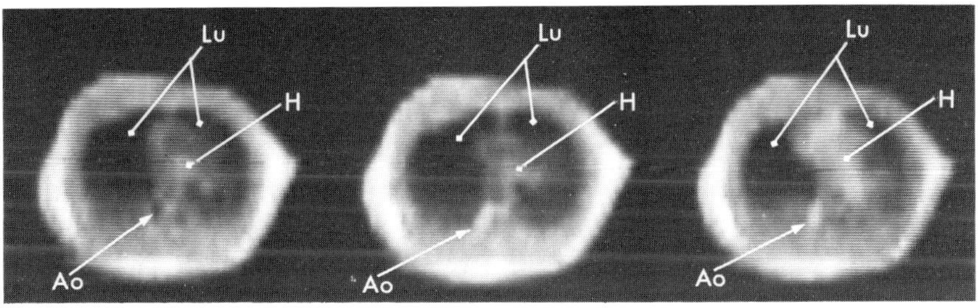

Figure 5.24. *Left*: Ungated image through the chest of a live rat. Heart (H) and aorta (Ao) show a low intensity similar to that of lungs (Lu). *Right:* The same slice obtained after death. Both the heart and aorta yield intense signals because blood flow has ceased. An interesting opportunity to observe the immediacy of the change in the image presented itself in this animal: Because of the combined effects of a surgically induced myocardial infarct and anesthesia, the animal died during the scan shown at center. Just before the start of imaging (which goes line by line from top to bottom in 4 min), thallium 201 was injected. The animal was dead at the end of the imaging procedure, but a subsequent radionuclide image showed that the radionuclide had sufficient time to distribute normally; in other words, the animal had lived for at least 1 or 2 min after the injection. This finding is consistent with the observation that the top half of the center image matches the image of the live rat on the left and that the bottom half resembles the image of the dead rat on the right.

At a certain speed (typically, at velocities of 5–10 cm/sec), nuclei are traveling fast enough that none produce a signal; thus, a completely dark region is found. This region has been observed in large vessels in the rat, where the effect is well demonstrated in transverse sections through the chest of a rat that died during imaging (Fig. 5.24). While the rat is alive, the aorta is dark and the moving mediastinal contents are ill-defined, but after the rat dies, the aorta becomes bright because of the static blood, and the mediastinal contents are better defined. The same effect can be observed in NMR images through the brain of a rat both before and after death: The stapedial arteries are seen as dark spots in the image of the live rat, but they become more intense after the rat dies and blood flow has ceased (Fig. 5.25).

The ability to observe the flow of blood without the need for invasive procedures or contrast agents is exciting. In recent studies in phantoms, we have measured NMR intensity as a function of the speed of the flowing media and observed its distributions across a section of a tube. "Ring" regions of NMR signal occur on the inside walls of tubes, where, because of boundary layer effects, flow is slower than in the center of the tubes (Fig. 5.26). Obstructions can also be observed as regions of increased intensity (Fig. 5.27).

An interesting effect occurs at low speeds. Because fluid entering the volume has not previously undergone NMR irradiation, its state of polarization is maximal. Nuclei flowing into the imaging volume produce strong signals that compensate for the loss of signals from the nuclei leaving the volume. As a consequence, at low speeds we observe a paradoxical enhancement, as shown in Figure 5.28, in which laminar flow produces a speed gradient that is seen as a gradual increase in NMR intensities from the center of the phantom toward the walls. This effect, which occurs at speeds of a few centimeters per second, may be important in enhancing the intensity of signals from blood in nearly occluded vessels.

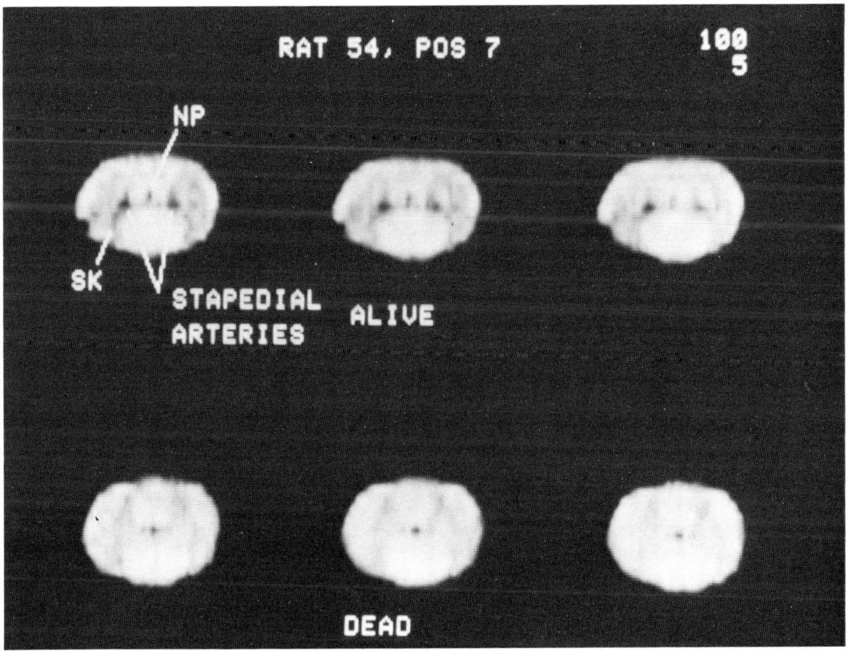

Figure 5.25. Image through the head of a rat. Top row, alive; bottom row, dead. From right to left, b = 0.25, 0.5, and 1.0 sec, respectively. While the rat is alive, the stapedial arteries (a branch of the internal carotid artery in rats) are dark due to blood flow. After blood flow stops, the arteries are no longer perceptible. NP, nasopharynx; SK, skull.

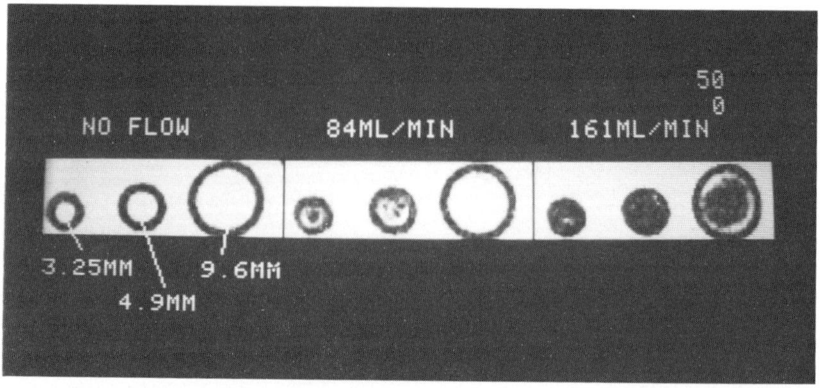

Figure 5.26. NMR cross-sectional images of a phantom containing three plastic tubes with inside diameters as indicated. The qualitative relationship between flow and NMR intensity is made evident by the "rings" produced in regions of the tube where, because of boundary-layer effects, fluid is moving at a slower velocity.

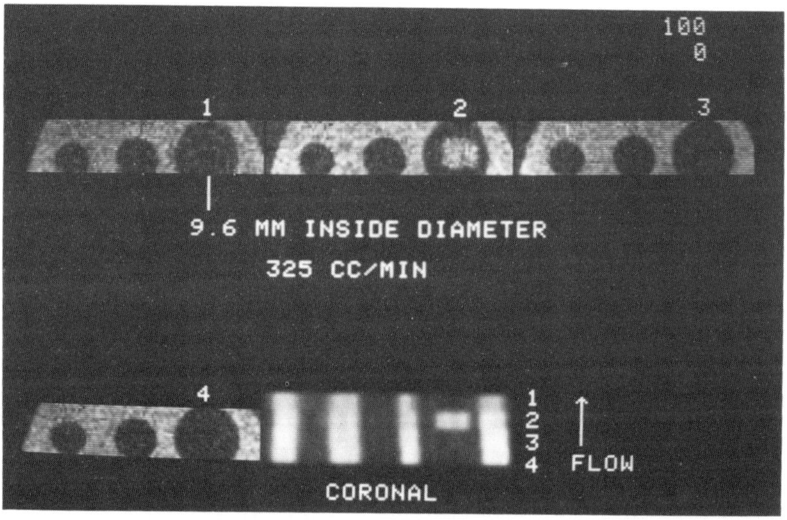

Figure 5.27. Four adjacent NMR sections of the same phantom shown in Figure 5.26. An obstruction containing stationary fluid is clearly evidenced in three transverse sections and in an image that formats the data into a coronal view.

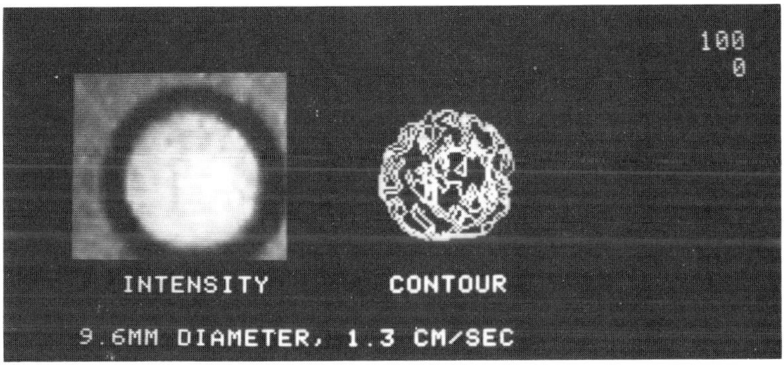

Figure 5.28. Paradoxical enhancement is evidenced in the NMR intensity image of a 9.6-mm inside diameter tube through which fluid flows at a velocity of 1.3 cm/sec. Note the brighter appearance of the moving fluid compared to that of the stationary fluid outside the tube. Contour map shows intensity increasing from the rim of the tube (where flow is at a lower velocity) to its center, where velocity is highest.

GENERAL CONSIDERATIONS

A dramatic aspect of NMR imaging is the high intrinsic object contrast it provides while at the same time maintaining good spatial resolution. The relative intensity scale is different from those previously encountered in diagnostic medicine and does not correlate directly with any physical or physiologic phenomenon previously used for diagnosis. The gray scale for NMR is different for different imagers and even for different settings of the same imager. Nevertheless, a pattern is emerging whereby NMR imaging, without the use of contrast agents, demonstrates sensitivities and specificities that exceed those found under similar conditions in other imaging modalities. On the other hand, the technique does not appear capable of realizing "automatic" detection of disease, and trained professionals will need to integrate the information that NMR yields with other data.

NOTE

Many of the photographs in this chapter have previously appeared in references 1 and 3, or will soon appear in references 13 and 14.

REFERENCES

1. Crooks LE, Hoenninger JC, Arakawa M, et al: Tomography of hydrogen with nuclear magnetic resonance. *Radiology* 136:701–706, 1980.

2. Crooks LE: Selective irradiation line scan techniques for NMR imaging. *IEEE Trans Nucl Sci* NS-27:1239–1244, 1980.
3. Hansen G, Crooks LE, Davis PL, et al: In vivo imaging of the rat with nuclear magnetic resonance. *Radiology* 136:695–700, 1980.
4. Damadian R: Tumor detection by NMR. *Science* 171:1151–1153, 1971.
5. Damadian R, Zaner K, Hor D, et al: Human tumors detected by NMR. *Proc Nat Acad Sci USA* 71:1471–1473, 1974.
6. Medina D, Hazlewood CF, Cleveland G, et al: NMR studies on human breast dysplasias and neoplasms. *J Nat Cancer Inst* 54:813–818, 1975.
7. Eggleston JC, Saryan LA, Czeisler JL, et al: NMR studies of several experimental and human malignant tumors. *Cancer Res* 33:2156–2160, 1973.
8. Hazlewood CF, Cleveland G, Medina D: Relationship between hydration and proton NMR relaxation times in tissues of tumor bearing and non-tumor bearing mice. *J Nat Cancer Inst* 52:1844–1853, 1974
9. Inch WR, McCredie JA, Knispel RR, et al: Water content and proton spin relaxation times for malignant and non-malignant tissues from mice and humans. *J Nat Cancer Inst* 52:353–356, 1974.
10. Weisman ID, Bennett LH, Maxwell LR, et al: Recognition of cancer in vitro by NMR. *Science* 178:1288–1290, 1972.
11. Hollis DP, Saryan LA, Eggleston JC, et al: NMR studies of cancer. *J Nat Cancer Inst* 54:1469–1472, 1975.
12. Barriolhet LE, Moran PR: NMR relaxation spectroscopy in tissues. *Med Phys* 2:191–194, 1975.
13. Herfkens R, Davis PL, Crooks LE, et al: NMR imaging of the abnormal live rat and correlation with tissue characteristics. *Radiology* (In press).
14. Davis PL, Kaufman L, Crooks LE, et al: Detectability of hepatomas in rat livers by nuclear magnetic resonance imaging. *Invest Radiol* (In press).

6

Human NMR Imaging

James M. S. Hutchison, B.Sc., Ph.D.

Francis W. Smith, M.B.Ch.B., D.M.R.D., F.F.R.

The basic principles of NMR imaging have been described in Chapter 2 and, in particular, the procedure of Selective Excitation has been discussed in Chapter 3. In this chapter, we shall describe the particular techniques and instrumentation used in the Aberdeen whole-body imager and present some results of clinical trials carried out on volunteer patients at Aberdeen Royal Infirmary during the latter half of 1980.

The imager was designed to produce transverse sectional images at any level on the body and to display two different kinds of image from each section, namely, one representing proton density and another representing spin-lattice relaxation time, T_1. Furthermore, the image processor allows the operator to make detailed spot and average measurements of T_1 over selected regions of the image. We feel that this ability to extract numeric information is invaluable in clinical reporting, and it is discussed further in the sections on Image Characteristics and on Images of Healthy and Ill Volunteers.

SPIN-WARP IMAGING

Imaging Techniques

Many different techniques exist for acquiring data to form NMR images. One way of classifying these techniques is based on shape of the "active region" of the sample, that is, the region that is effectively contributing to the NMR signal at any one time. The active region can approximate a point, a line, a plane, or the total volume. As examples of each technique, we have the original sensitive-point method of Hinshaw (1) and FONAR (2), which are point methods. Line methods include those of Mansfield et al. (3) and Hutchison et al. (4). Planar methods include most tomographic techniques that use projection reconstruction (5,6), and Lauterbur and Lai (7) have produced some interesting three-dimensional images using projections distributed over a sphere, representing the total-volume category.

In principle, a total-volume method is the most efficient because all the spins contribute to the NMR signal during each signal collection period, but the practical problems are severe. Among these problems are the requirements of a very homogeneous main magnetic field, accurate and linear gradient fields, lack of subject movement, and long total data-acquisition times. At the other end of the scale, point methods are quite tolerant of field nonuniformities and subject movement, but the rate of data acquisition is low.

Our whole-body imager was originally designed to use a line method and gave reasonable results on static phantoms but was less successful on live human subjects. Although tolerant of field nonuniformity, it turned out to be highly intolerant of subject movement and of field instability. Unfortunately, these two effects will always be present in a clinical environment: Patients must breathe, and their hearts must beat! The field stability and uniformity problem is not so obvious to the pure physicist working in a magnetically shielded, isolated laboratory. In the clinical environment, many unpredictable factors can combine to alter a magnetic field pattern. Part of the protocol must be to divest patients of all ferromagnetic items (including watches and safety pins), but there is a limit on how far this protocol can extend to the attendant staff, and many items of everyday clothing contain ferromagnetic parts. Even outside the immediate environment of the magnet, ferromagnetic objects can disturb its field: The arrival of a wheelchair within a few yards can upset the magnetic field pattern.

The conversion of our imaging technique to the planar method, in fact, necessitated little hardware modification but resulted in a much improved image requiring only one fourth of the data-acquisition time. We have given the name "spin-warp imaging" to our technique, which is described below and has also been described by Edelstein et al. (8).

Spin-Warp Imaging

The pulse sequence used in this method is shown in Figure 6.1, whose horizontal axis is time. The main magnetic field **H**, acts vertically in the z direction. The y axis runs horizontally along the long axis of the patient, and x runs horizontally across the patient. The aim is to produce a tomographic image that corresponds to the x-y plane (y = 0).

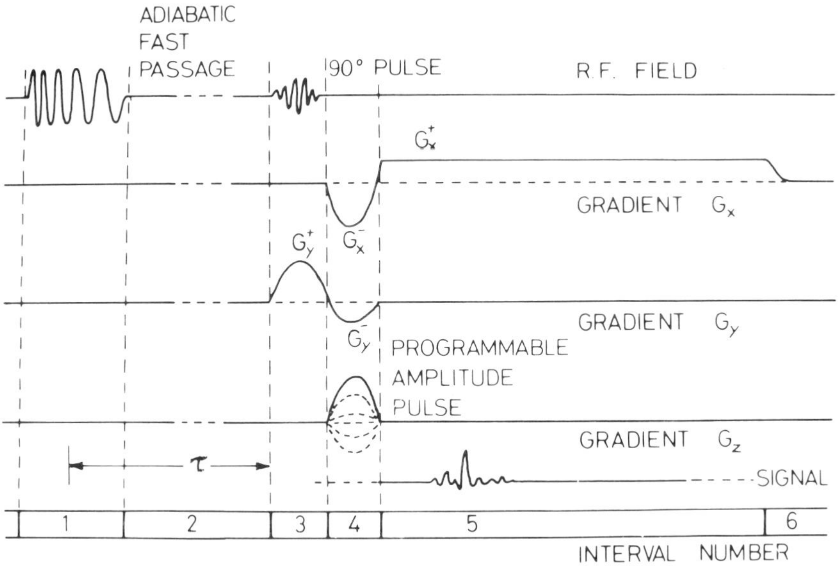

Figure 6.1. Pulse-sequence spin-warp imaging.

Consider first the events beginning with interval 3. Here we apply a shaped 90-degree radio frequency pulse along with a field gradient in the y direction, G_y, which selectively excites the spins within a narrow slice in and close to the x-z plane.

During interval 4, all three orthogonal field gradients are applied. At first sight, this approach would appear to make analysis of the spin behavior rather difficult, but since no radio frequency field is present, the effects of the three gradients can be considered independently; the net effect is the same as if the three were performed in sequence, and the coincidence in timing is simple, a convenience that serves to save time between excitation and signal acquisition. G_y^- serves to rephase the spins across the width of the slice and hence maximize the signal finally obtained; the optimal conditions for this approach have been discussed by Hutchison et al. (9). G_x^- serves to dephase the spins along the x direction as a preliminary to the readout gradient G_x^+ in interval 5, which causes them to rephase to produce a kind of gradient-induced spin echo in the middle of the signal-acquisition period. The Fourier transformation of this spin echo is therefore a one-dimensional projection of the spin density within the slice onto the x axis.

A programmable amplitude pulse, G_z, is introduced during interval 4 to give a known amount of twist, or "warp," to each vertical column of spins and thus phase-encodes it before projection onto the x axis. In effect, G_z maximizes the response to a particular vertical spatial frequency in the column, equal to the spatial frequency of the "warp." The whole sequence is repeated several times, each time using a different size of G_z pulse, and hence covering a range of vertical spatial frequencies from zero up to some maximum. If the projected spin-density values for any one column (obtained from the transformations of the spin-echo signals) are arranged in order of increasing G_z pulse size and subjected to another Fourier transformation, this sequence will represent the distribution of spin density up the column (z). If we do this sequence for each column, we have a complete two-dimensional image of our selected slice. Thus, as $N \times N$ image can be obtained by taking N projections into the same axis, a procedure that is obviously impossible in x-ray or radioisotope imaging, simply because the basic signal has no phase information in it.

The spin-warp technique is similar, in principle, to Fourier zeugmatography, as described by Kumar et al. (10). However, it differs in one important aspect, namely, that in our method, the time between the 90-degree pulse and signal data acquisition is the same for all sequences, whereas in the Fourier zeugmatography method it is not—it varies by the length of one gradient pulse and not its amplitude. One major consequence of this difference is that our method is much less sensitive to inhomogeneity of the main magnetic field.

Spin-Lattice Relaxation Time

By including the events of intervals 1 and 2 in Figure 6.1, we can obtain information about T_1. All the spins in the region of influence of the radio frequency coil are inverted during interval 1, then allowed to relax partially during interval 2; at interval 3, we commence the sequence to collect signal data exactly as before. If we call these signals S_2, while those derived without prior inversion of the spins are called S_1, for each pixel in the final image, we can compute:

$$T_1 = \frac{-\tau}{\ln\left\{\dfrac{|S_1 - S_2|}{|S_1|}\right\}},$$

where τ is the time from effective inversion of the spins to the center of the 90-degree pulse (190 msec in our case). Notice that S_1 and S_2 are subtracted before discarding the phase information because S_2 can have the opposite phase from S_1 if T_1 is long (>270 msec). Also, in practice, it is necessary to cut off the algorithm if $|S_1|$ falls below a certain value; otherwise, noise becomes dominant and produces false T_1 values. In the areas where the algorithm has been cut off, the displayed image is set to black ($T_1 = 0$, in effect).

In our case, spin inversion is achieved using adiabatic fast passage (AFP). This technique was used in the early days of NMR with continuous-wave spectrometers because it enabled one to invert spins without requiring the high-peak radio frequency powers commonly used in the more familiar 180-degree pulse. It is this feature that makes this technique attractive for whole-body imaging.

To perform AFP, a radio frequency magnetic field is applied to the sample with an initial frequency somewhat below the Larmor precession frequency of the nuclei. Its frequency is then swept upward through resonance at a controlled rate until it reaches an equal offset on the other side of the Larmor frequency, when it is switched off. The effect on the spins is better understood by considering them in the rotating frame, that is, a coordinate system rotating with the spins at the Larmor frequency around the main field ($\mathbf{H}_0$) direction as an axis. In this frame, the effective magnetic field is seen to start off approximately aligned with the nuclear magnetization vector; it then reduces in length and gradually rotates, passing the origin on one side, then lengthens again and ends up pointing almost exactly opposite its starting point. If this process occurs smoothly enough to be adiabatic, the spins remain aligned with this field vector throughout and hence end up in the inverted state. The term "fast" implies that the procedure must be completed in a time that is short compared with either T_1 or T_2 (the spin-spin relaxation time).

When compared with the 180-degree pulse method, inversion by AFP is much less affected by inhomogeneities in either the main field, $\mathbf{H}_0$, or in the radio frequency field. It also gives efficient inversion over a region much wider than the imaging plane. These advantages are reflected in the T_1 images as a lack of dependence of displayed T_1 on position within the image plane and insensitivity to flow in the case of blood (see section on Image Characteristics). AFP takes somewhat longer to perform (10 msec) than a 180-degree pulse but is still quite short compared to the relaxation interval (190 msec).

APPARATUS

Figure 6.2 is a photograph of the apparatus before completion of the screened enclosure. One of us (J.M.S.H.) is in position for imaging. The magnet is a four-coil, air-cored system manufactured by Oxford Instruments, Ltd. of England and produces a field of 0.04 Tesla (400 G) when energized at 40 A, leading to a proton resonance frequency of 1.7 MHz.

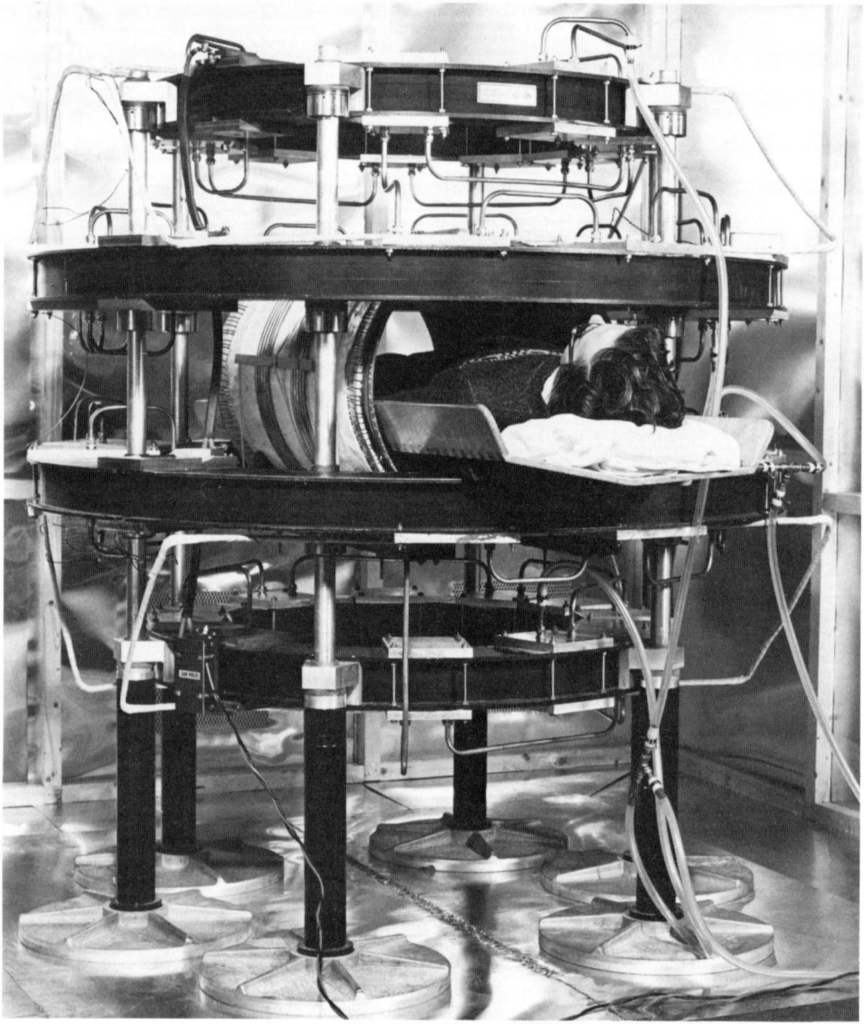

Figure 6.2. Four-coil air-core imaging system.

The two large coils have an average diameter of 1.3 m and are separated by 0.3 m, allowing a sufficient gap for most patients to be inserted. The homogeneity was adjusted to within ± 3 in 10^4 over a diameter of 0.46 m in the imaging plane.

There are three sets of field gradient coils. The y gradient set, seen in Figure 6.2 as four white rectangles, produces a peak field gradient of 2.7 m Tesla/m (0.27 G/cm) along the length of the patient (y). The other two gradients are produced by windings distributed on the plastic cylindric former surrounding the patient. These windings consist of straight wires running axially along the cylinder, with winding densities proportional to cos 2θ and sin 2θ, respectively. Each winding has 19 wires with a diameter of 1.8 mm in each quadrant, and the windings are interlaced in such a way that no two adjacent wires are closer than twice their diameter. They also act as a Faraday shield for the radio frequency coil wound on the outside of the former (11).

The radio frequency coil consists of two windings, each of five turns of 8 mm copper tubing, connected in parallel and operated in an electrically balanced mode. This configuration, along with the shielding provided by the gradient windings, gives virtually no electric radio frequency field inside the cylinder. The cylinder diameter is 500 mm, and the unloaded Q factor of the radio frequency coil is about 400, dropping to about 200 with a patient inside.

The screened room is a cube 2.4 m on each side, lined with aluminum sheeting 0.46 mm thick; all sheets are electrically connected by being spot-soldered at 25-mm intervals. This completion of the electrical conducting circuit is vital to screen out interfering radio frequency magnetic fields. A standard-size door, fitted in one side of the screened cube to allow patient and operator access, is also lined with aluminum sheet and surrounded by copper-spring draught excluder to make electrical contact when closed. The power connections through the screening for the magnet, gradient windings, and internal light are all filtered to remove radio frequency interference. The patient lies on an entirely nonmetallic couch that slices through the sensitive region of the apparatus on polytetrafluoroethylene (PTFE) runners (not shown in Fig. 6.2).

To collect the data for one image section, 64 sequences without and 64 sequences with AFP inversion are carried out. These sequences are alternated once per second. Sixty-four different sizes of G_z pulse (from -32 to $+31$ times the unit size) are used, so the whole process takes 128 sec. Once the apparatus has been set up, data collection can be initiated from inside the screened room, and after 128 sec, a lamp indicates completion, so that the operator can move the patient to a new position and start accumulating the next set of data. One person can be in charge of the patient, of changing and selecting the section to be imaged, and of operating the apparatus without having to leave the screened room. The apparatus is interfaced to a PDP 11/34 computer through a custom-built analog-to-digital converter (12-bit, two-channel, simultaneous sampling) in which signal data are initially stored and then Fourier transformed and a T_1 matrix calculated. Each 64×64 image matrix is interpolated to 256×256 and displayed on a CONRAC display in both monochrome and color.

IMAGE CHARACTERISTICS

The sections presented in the remaining part of this chapter were obtained directly from the monochrome display, which most observers agree is a better medium than color for viewing proton-density images. However, the ability to color-code spin-lattice relaxation times is valuable to clinicians, enabling them to assess the site, nature, and extent of a lesion fairly quickly, as well as forming a basis for pattern recognition: The brain remembers colors better than shades of grey. The effective section thickness is 17 mm, as determined by the relative strengths of the radiofrequency field during the 90-degree pulse and the y gradient applied at that time. The spatial resolution is 7 × 7 mm and is interpolated in the display so that each display pixel is 1.75 × 1.75 mm.

The noise level on proton-density images is about 2.5% r.m.s. of the value corresponding to 100% water. On T_1 images, noise level depends on proton density—for copper sulfate solution, it is about 3.5% of T_1 near 200 msec, gradually worsening as one departs from this optimal value. The usefully measurable range of tissue T_1 values is from 100 to 500 msec, approximately and, on the basis of a tissue water level of 70%, the noise on T_1 images should never exceed 10% of T_1 itself. The Conrac imager display provides a joy-stick-operated marker that may be used to outline a region of interest and enable the computer to calculate the average T_1 of a region, along with its standard deviation as shown, for example, in Figure 6.3.

The sensitivity to proton density is not quite uniform over the slice and falls off fairly rapidly at the sides, where the patient's arms are normally placed. A first-order correction for left-right imbalance has been applied to all proton-density images, but it is immaterial for T_1 images, which are based on the ratio of two signals and not their absolute size. Apart from the standard windowing facility provided on the Conrac display, this image correction is the only one carried out: There is no smoothing, apodization, deletion, or geometry correction.

Geometry tests with a regular rectangular array of small bottles of copper sulfate solution give results consistent with theory, namely, that horizontal lines appear straight and vertical intervals are reproduced correctly. Vertical lines, on the other hand, appear curved, and horizontal intervals are not reproduced linearly. The magnitude of this geometric error depends directly on the main field nonuniformity and is predictable, if not easily correctable. There is, however, no smearing of the image, as would certainly occur if multiple-angle projection reconstruction were used under the same conditions. To date, no problems of clinical assessment have arisen as a result of the small amount of distortion actually present.

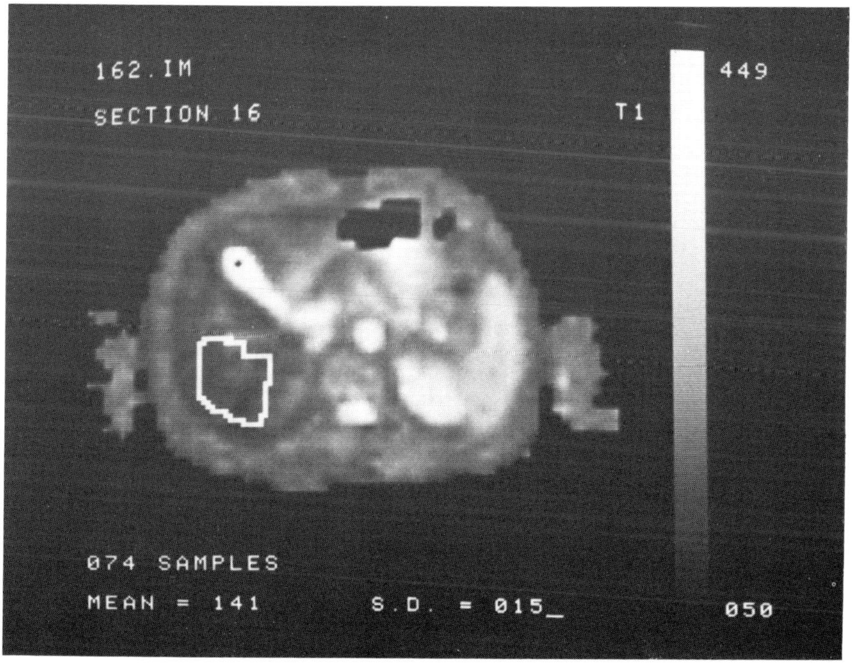

Figure 6.3. T_1 image of human abdomen. The region of interest in the liver has a T_1 value of 141 msec.

Normal involuntary movements, such as breathing and heart function, do not seem to impair images. The two ventricles of the heart are regularly seen quite clearly (e.g., Fig. 6.4), and the T_1 of the blood in them seems to measure correctly, despite its considerable motion. The T_1 of blood varies in the range 340 to 390 msec. Because the blood that is in the imaging plane could well have been some distance outside it when its spins were inverted, this shows the usefulness of AFP as a wide-region, spin-inverting procedure when T_1 measurements are required. Restlessness, which causes a patient to assume a slightly different posture part way through a section, can produce major image artifacts. These artifacts are usually characterized by areas of apparently positive proton density outside the normal boundary of the section, which we jokingly call "ectoplasm," and are quite different from those caused by ferromagnetic objects on or around the patient's body. These metallic artifacts appear as a crescent-shaped absence of image on the periphery of the section caused by distortion of the magnetic fields. An example is shown in Figure 6.5, which was caused by an unsuspected safety pin.

Another oddity of the imaging process is commonly observed in regions

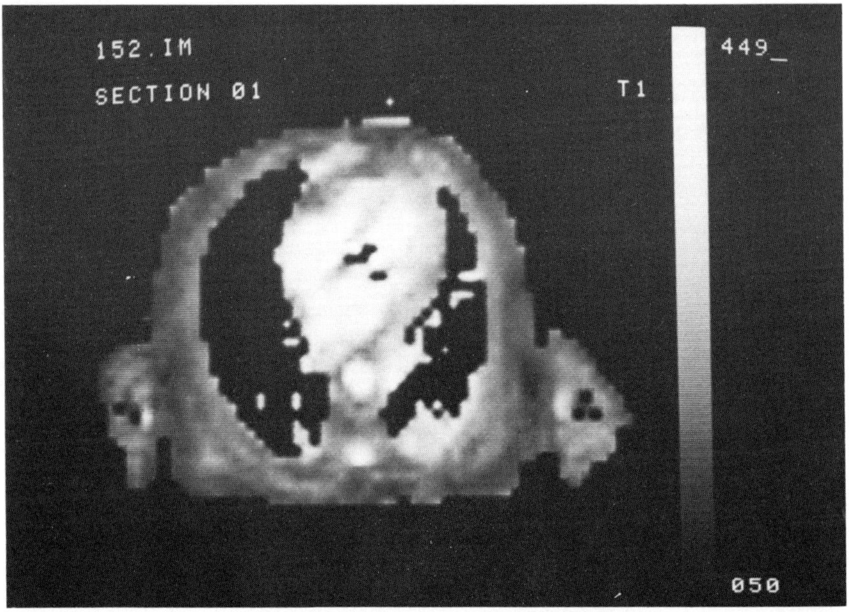

Figure 6.4. (legend on facing page)

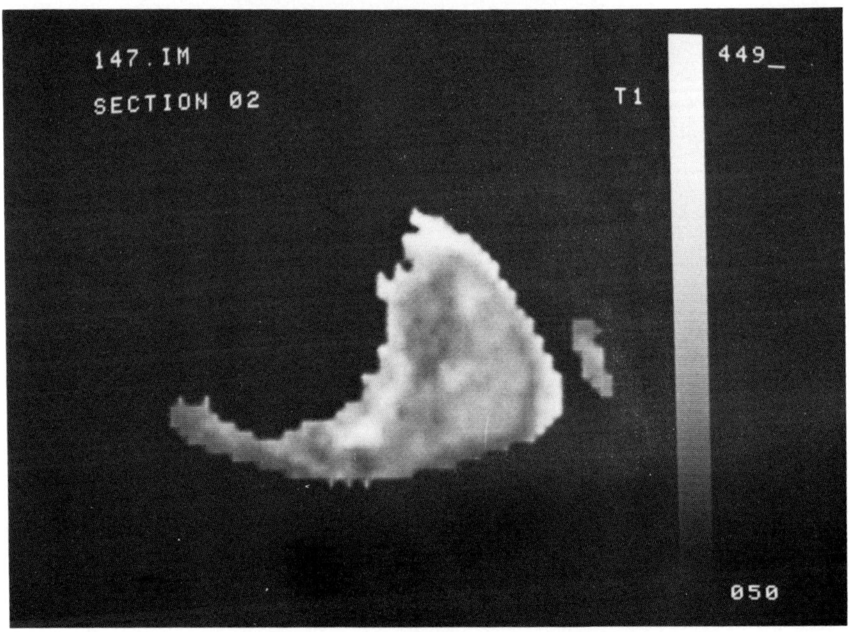

Figure 6.5. (legend on facing page)

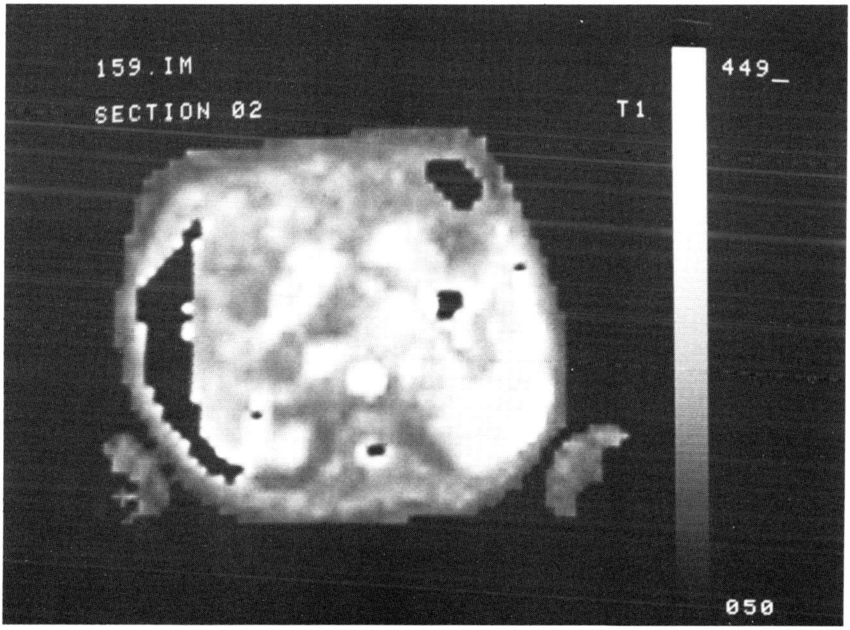

Figure 6.6. The upper abdomen of a subject with ascites. Dark areas in the T_1 image illustrate the appearance of fluids with long relaxation times.

that contain fluids with long relaxation times. In these regions, the proton spins do not recover sufficiently in the 1-sec interval between pulses, leading to an apparent reduction in proton density, sometimes to the extent that the T_1 algorithm cuts off, and the area registers as black. Typically this effect happens with urine in the bladder and some cysts and ascites as, for example, in Figure 6.6.

IMAGES OF HEALTHY AND ILL VOLUNTEERS

In this section, we shall present NMR transverse sectional images of both healthy persons and those with a variety of pathologic conditions.

Figure 6.4. Thoracic section taken at the level of the tricuspid valve; the two ventricles are seen quite clearly despite normal cardiac motion.

Figure 6.5. Image with crescent-shaped artifact caused by the distortion of the magnetic field due to the presence of ferromagnetic objects on or around the subject.

Head

Figures 6.7a and 6.7b are sections through the cerebral hemispheres at the level of the lateral ventricles, showing on the T_1 image (Fig. 6.7a) an irregular outline comprising skin, muscle, and bone. The grey and white matter are clearly seen, as are the large ventricles. The proton-density image (Fig. 6.7b) shows more clearly the bone of the skull vault, overlain by skin and muscle, but the grey and white matter are not differentiated; the cerebral cortex appears as a patchy image, although the ventricles are more clearly demonstrated. The T_1 images are therefore of more use in studying the cerebral cortex, the proton-density images being referred to when looking for bone pathology or when differentiation of a long T_1 lesion is needed. Because cerebral edema has a T_1 in the range of 330 to 370 msec, which is shorter than cerebrospinal fluid (400-450 msec) and much longer than normal cerebral tissue (250-300 msec), it is useful to refer to the proton-density images to get a clear demarcation of the ventricular system, basal cisterns, and any cerebrospinal fluid in either the insula of the temporal lobe or the sulci of the cortex, in an effort to differentiate small areas of cerebral edema from cerebrospinal fluid.

Figure 6.8 is a normal-section T_1 image through the cerebral hemispheres demonstrating grey and white matter and falx cerebri and superior sagittal sinus.

Figure 6.9 is a T_1 image at the level of the third ventricle in a patient with cerebral atrophy. The third and lateral ventricles are enlarged, and cerebrospinal fluid is seen in the insula of both temporal lobes.

Figure 6.10 is a T_1 section through the base of the skull in a patient with a normal cerebellum and brain stem surrounded by white (long T_1) cerebrospinal fluid in the basal cisterns. The petrous and occipital bones appear black because they are below range, having both a very low mobile proton density and a short T_1 relaxation time. Similarly, the right maxillary antrum and sphenoid sinus appear black because they are filled with air. The left maxillary antrum is white, having a T_1 of 380 msec, and demonstrates the presence of an infected sinus with edematous mucosa.

Cerebral metastases, gliomas, and meningiomas that produce edema are not easily visible. The surrounding edema is clearly seen, as is any displacement of the ventricular system. Figure 6.11 demonstrates the edema surrounding a right occipital meningioma; it also shows that the left lateral ventricle has been flattened out and the right one displaced later-

Figure 6.7a. T_1 coronal section of the cerebral hemispheres.

Figure 6.7b. Proton-density coronal section of the cerebral hemispheres.

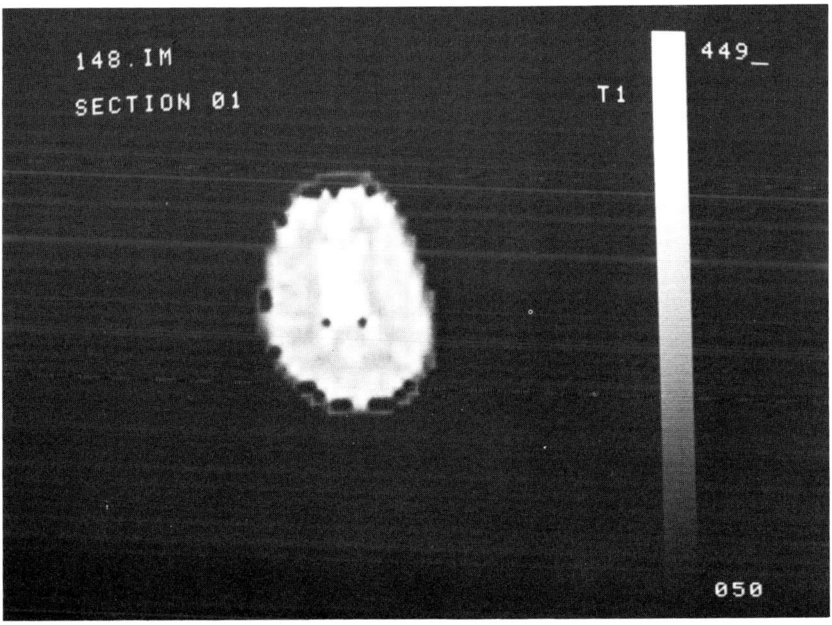

Figure 6.7a. (legend on facing page)

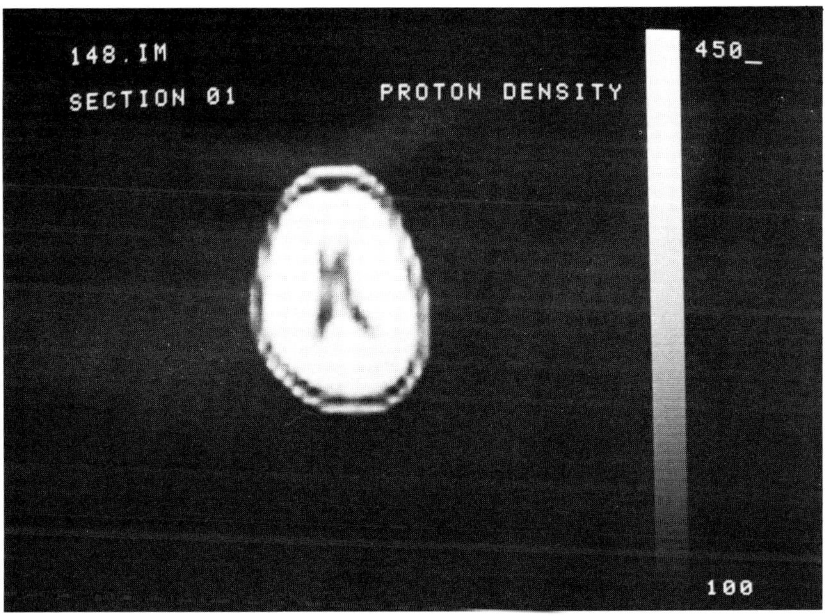

Figure 6.7b. (legend on facing page)

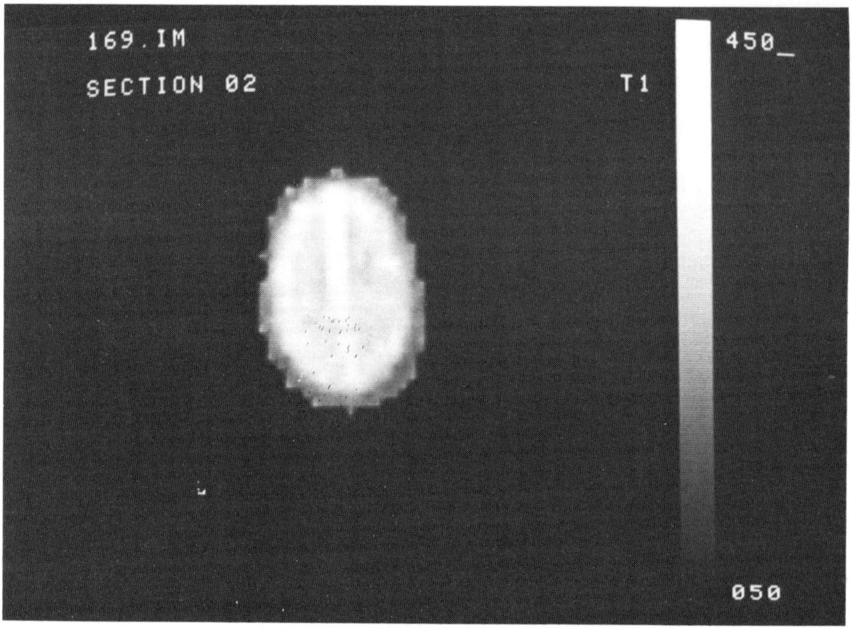

Figure 6.8. Normal T_1 image through the cerebral hemispheres.

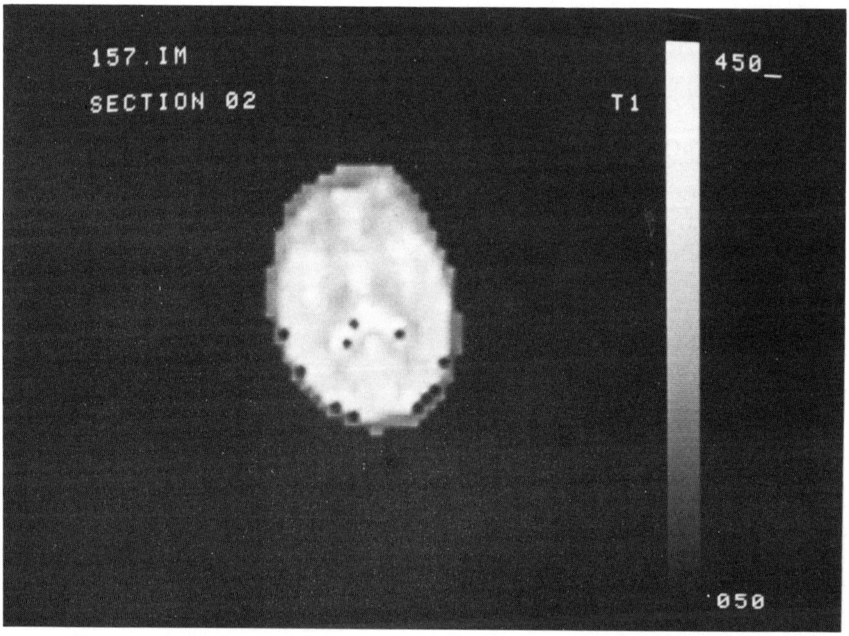

Figure 6.9. T_1 image of the brain at the level of the third ventricle.

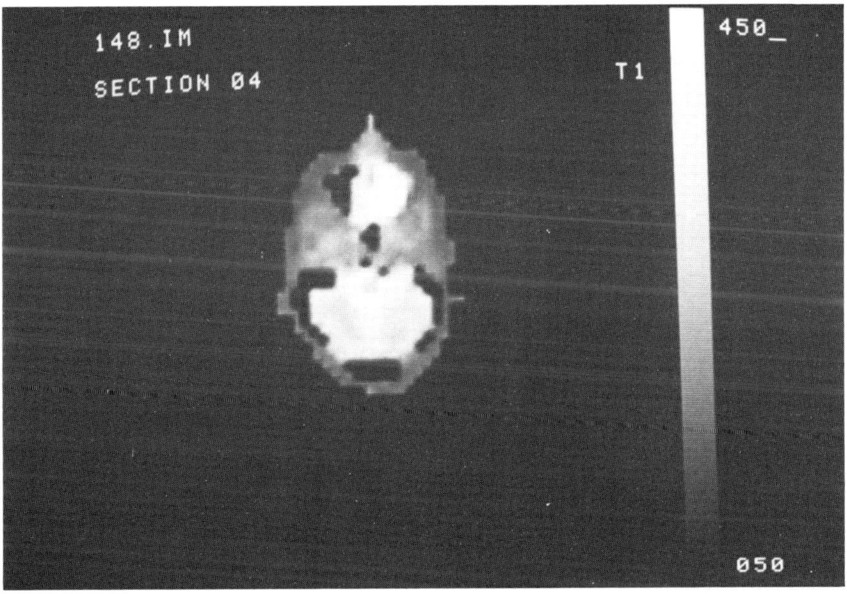

Figure 6.10. T_1 image through the base of the skull of a patient with a normal cerebellum and brainstem. The infected left maxillary antrum has a high T_1 time.

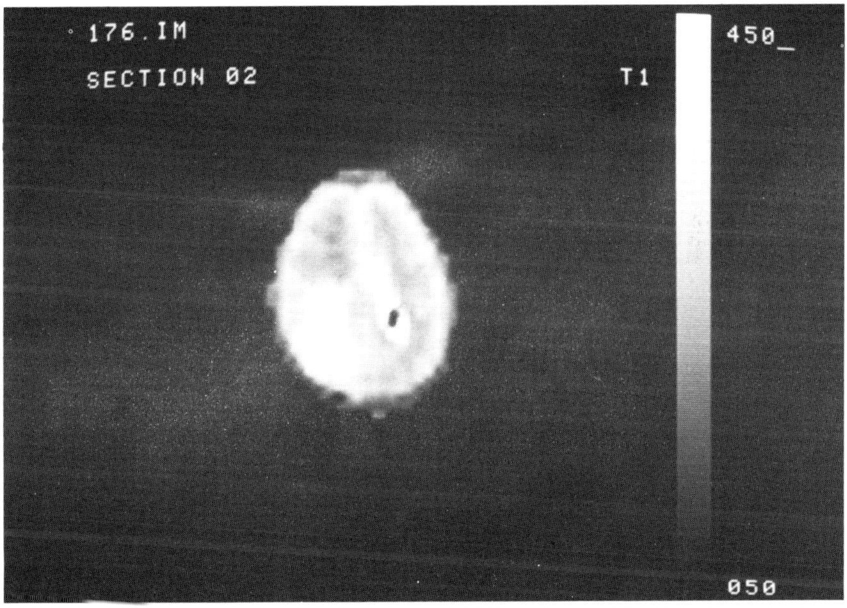

Figure 6.11. Edema surrounding a right occipital meningioma.

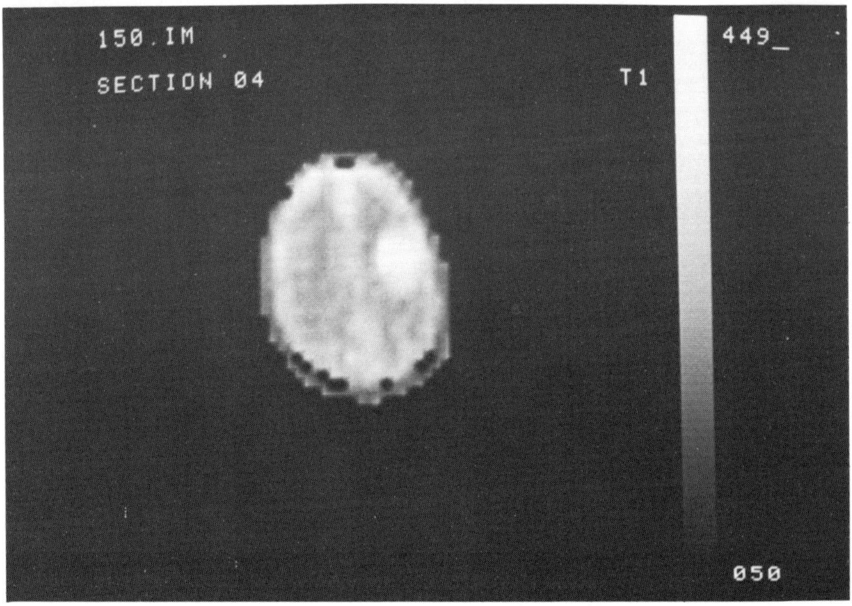

Figure 6.12. Cystic gliomas without edema.

ally. By contrast, the cystic glioma shown in Figure 6.12 has not produced any edema and is clearly seen. The small tumor seen at 2 o'clock as a small grey mass has a T_1 of 350 msec; the associated large cyst has a fluid level in it that was not seen on conventional x-ray CT. The precipitate in the cyst had a T_1 that was the same as that of the tumor (350 msec); the supernatant, a T_1 of 420 msec.

Thorax

In the investigation of thoracic contents, the T_1 images are, again, the most useful, proton-density images being of value in assessing the vertebral bodies and spinal canal. Figures 6.13a and 6.13b show both the T_1 (Fig. 6.13a) and proton-density (Fig. 6.13b) images at the level of the aortic arch. At this level, the aorta, trachea, and some pulmonary vessels are clearly seen on the T_1 images. In both the T_1 and proton-density images, the lungs appear black because, being filled with air, they give off virtually no signal. The proton-density image shows clearly the vertebral body and both arms, the humeri being clearly seen.

Figure 6.14, a T_1 image at about 2 cm below the level of Figure 6.13, shows the right and left main bronchi, the ascending and descending

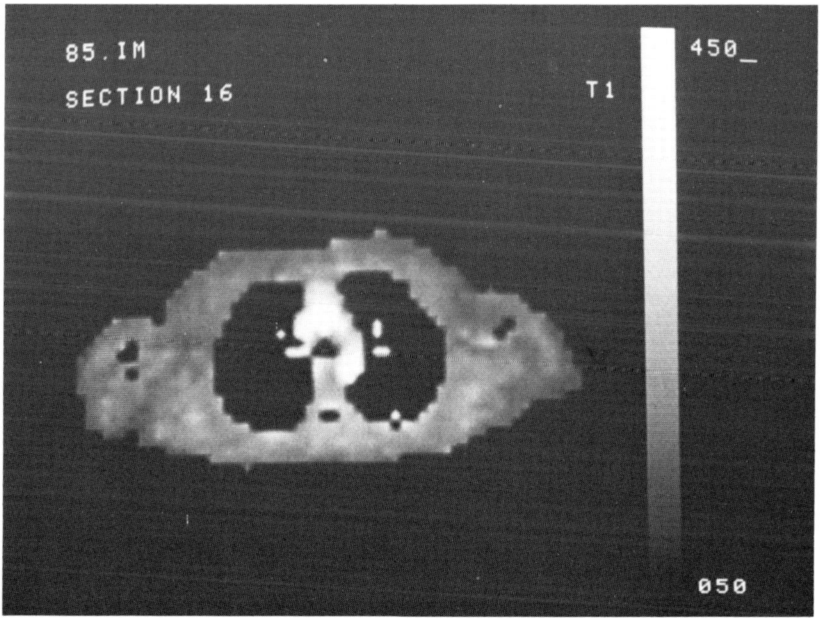

Figure 6.13a. T_1 image of the thorax at the level of the aortic arch.

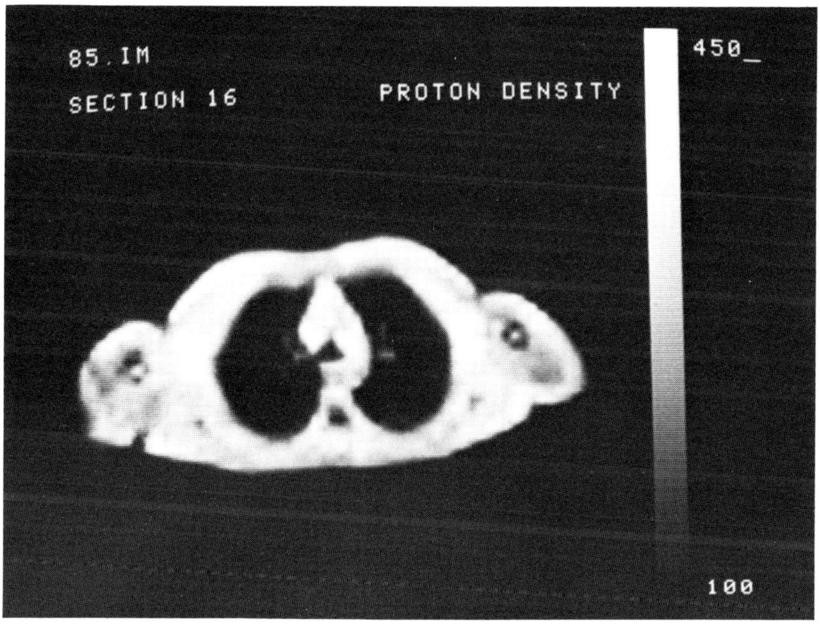

Figure 6.13b. Proton-density image at the level of the aortic arch.

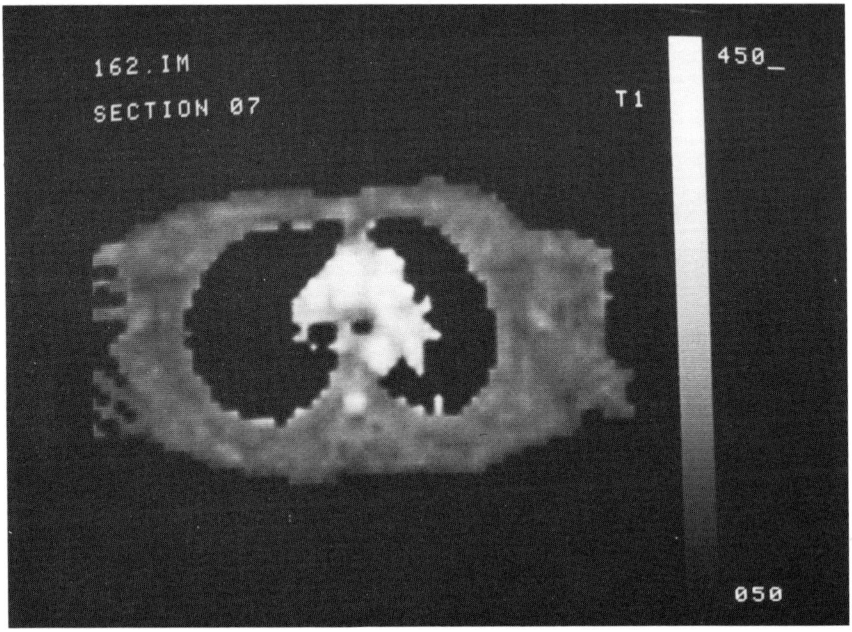

Figure 6.14. T_1 image, taken 2 cm below that in Figure 6.13, showing right and left main bronchi.

aorta, and the pulmonary arteries. Cerebrospinal fluid within the spinal canal appears white in the midline posteriorly.

Figure 6.4, taken at the level of the tricuspid value, shows the right atrium and ventricle and the left ventricle. The pulmonary vein and left atrium are seen lying behind the right atrium and in front of the descending aorta. The esophagus is not visible but lies in the dark grey area behind the pulmonary vein. The irregular mass lying in the angle of the rib on the left represents the swelling and reaction around a metastatic deposit from a renal carcinoma.

Figure 6.15 is a T_1 image taken at the level where the left pulmonary vein joins the main pulmonary vein and enters the left atrium. The right atrium and ventricle are seen anterior to it. The left lung shows pneumonic consolidation of the posterior segment of the upper lobe and has a T_1 of 390 msec.

Tumors of the lung and mediastinum have a T_1 of about 270 msec, ranging from 250 to 280 msec. A section at the level of the tracheal bifurcation (Fig. 6.16) shows the ascending and descending aorta with a large mass in the anterior mediastinum. The mass with a T_1 of 270 msec

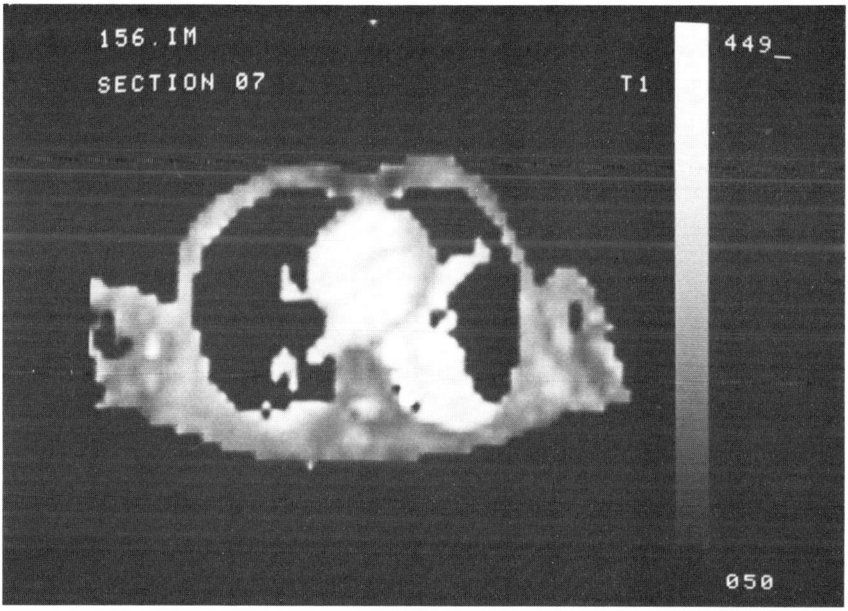

Figure 6.15. T_1 thoracic image taken at the level where the left pulmonary vein joins the main pulmonary vein.

had a center with a long T_1, about 500 msec, which is thought to represent the necrotic center to this squamous-cell carcinoma. Diagnosis was made by mediastinoscopy and biopsy.

Abdomen

Figures 6.17a and 6.17b show both the T_1 (Fig. 6.17a) and proton-density (Fig. 6.17b) images of a normal upper abdomen at the level of the first lumbar vertebra. The normal liver is seen on the T_1 images as an almost homogeneous grey with a T_1 in the narrow range of 140 to 170 msec. The gallbladder is seen clearly toward the portahepatis, and the bile duct is seen adjacent to the inferior vena cava. The abdominal aorta is anterior to the vertebral body of L_1, behind which is the white cerebrospinal fluid in the spinal canal. The upper pole of the left kidney is seen between the vertebral body and the spleen. The stomach is seen anterior to the aorta as a white mass over which there is an irregular black area representing the gastric air bubble.

In the proton-density image (Fig. 6.17b), there is a loss of definition between the liver and blood vessels, kidney, and spleen. The stomach,

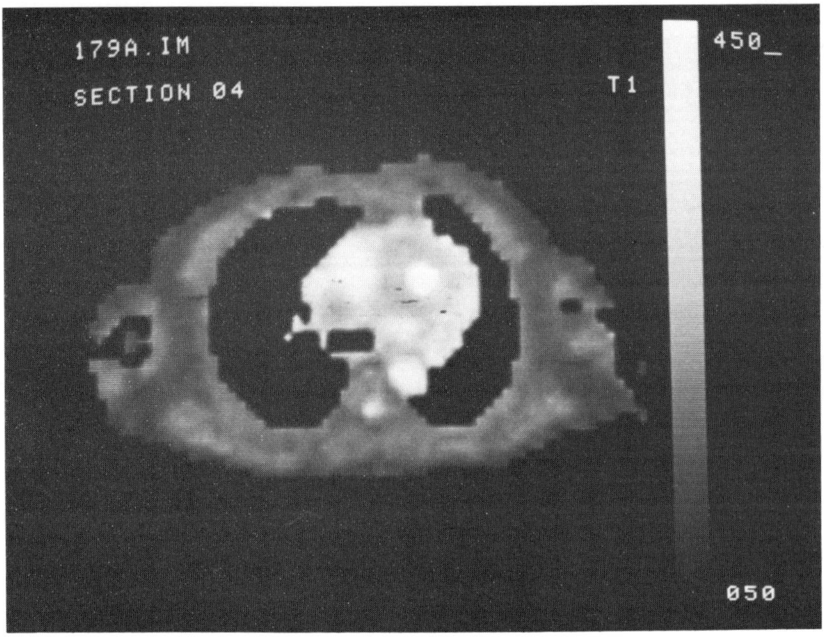

Figure 6.16. Thoracic T_1 image of a squamous cell carcinoma.

with its air bubble, and gallbladder are both clearly seen. Similarly, definition of the vertebral body is improved, the spinal canal and spinal cord being clearly demonstrated.

Figure 6.18 is a T_1 image at the level of the celiac axis. The celiac artery is seen arising from the aorta and dividing into the splenic and hepatic arteries. The portal vein is between the hepatic artery and the inferior vena cava. The liver appears normal, as do the kidneys and the spleen. The T_1 of the spleen has been found to be in the range of 250 to 290 msec. A small collection of gas in the descending colon is seen as an irregular black area anterior to the spleen. The bowel, both large and small portions, is not visible, except in cases of either fecal loading of the colon, or small-bowel ileus, or obstruction.

The pancreas has been seen on several occasions but not regularly, unless it is diseased; neoplasms show as a mass and pancreatitis as a uniformly enlarged pancreas with a longer T_1 than normal. Normal pancreas has a T_1 of 180 to 200 msec, whereas that with pancreatitis has a T_1 of 250 to 300 msec.

Figure 6.19, taken at the level of the renal vessels, shows both kidneys as horseshoe-shaped organs with a T_1 between 300 and 340 msec. Urine

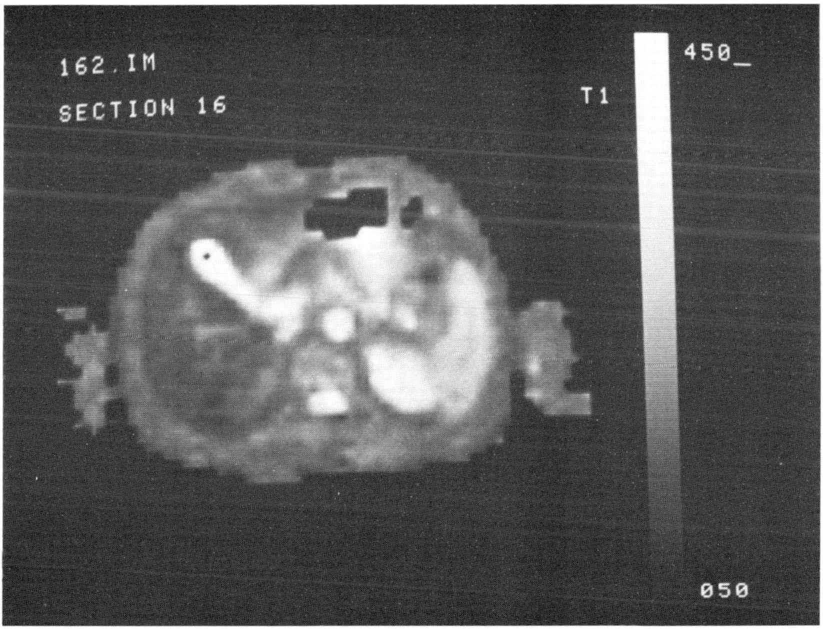

Figure 6.17a. T_1 image of a normal upper abdomen.

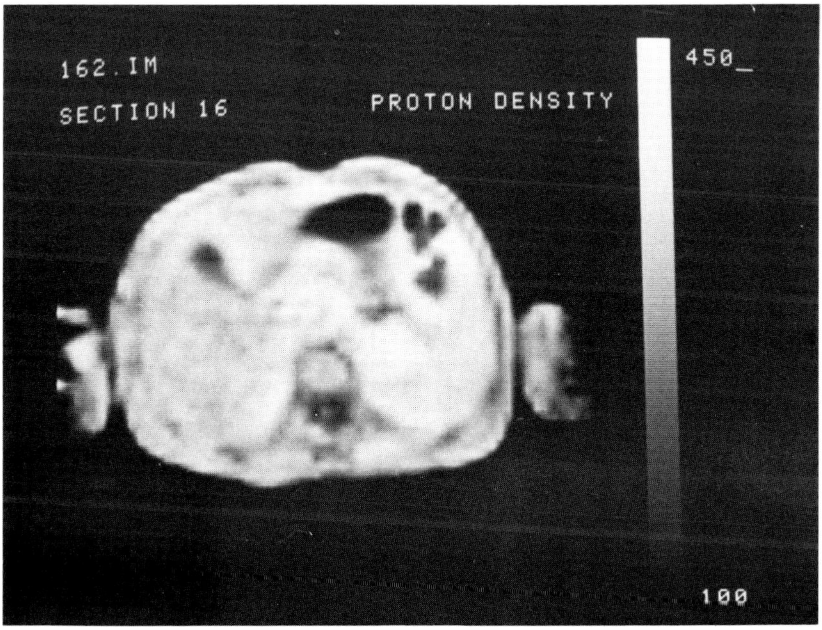

Figure 6.17b. Proton-density image of the same abdominal plane.

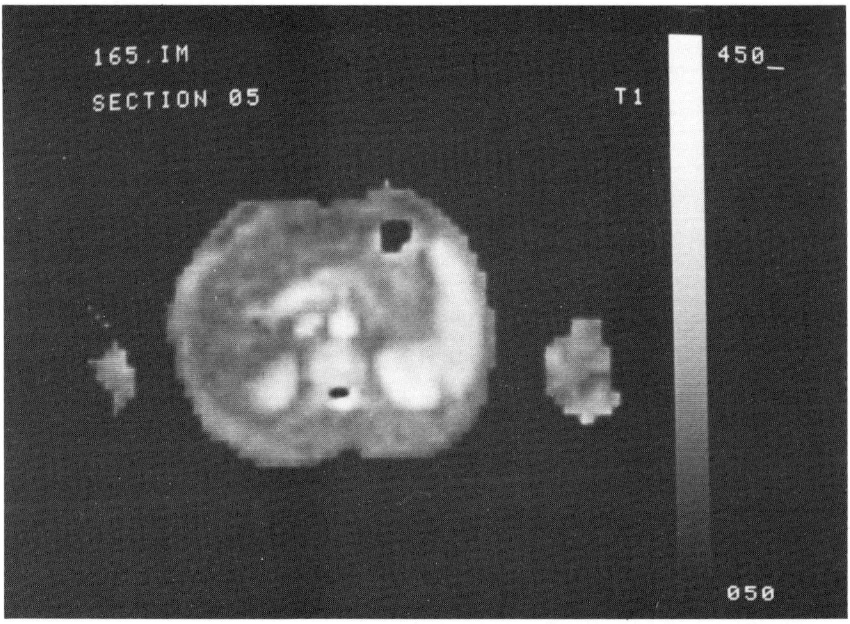

Figure 6.18. Normal T_1 image of the abdomen at the level of the celiac axis.

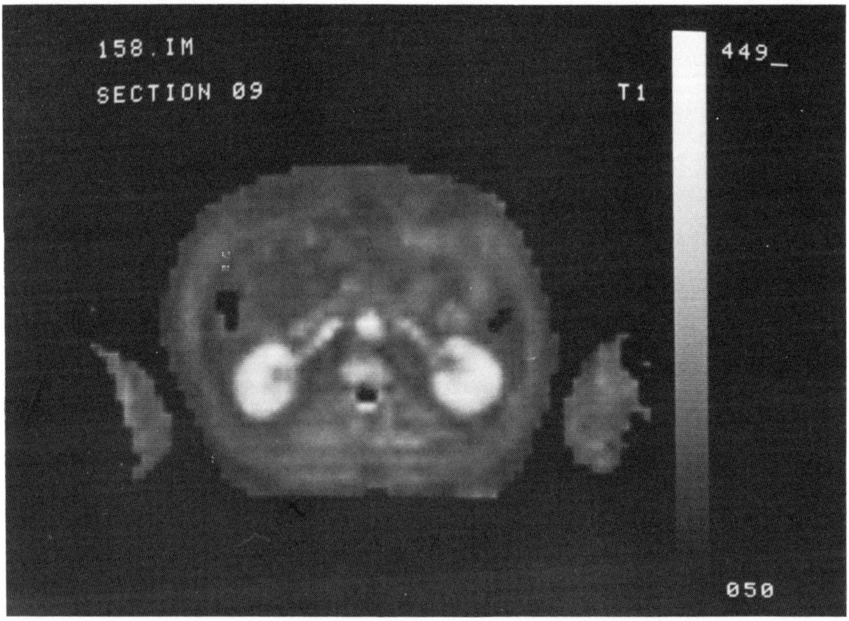

Figure 6.19. Abdominal image at the level of the renal vessels.

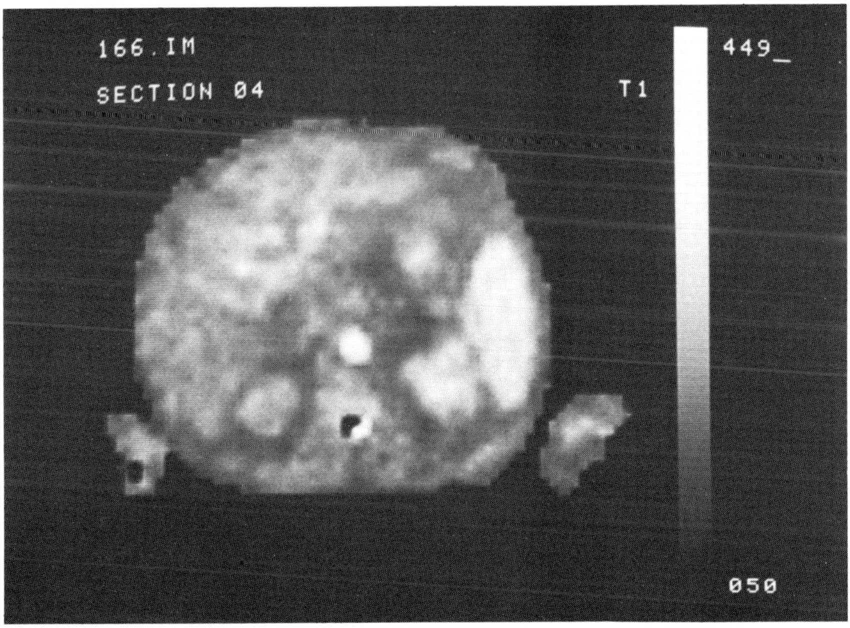

Figure 6.20. Cirrhotic liver.

within the urinary bladder has been shown in vitro to have a T_1 relaxation time of about 2 sec. The bladder, therefore, appears as a black structure with a T_1 in excess of 1 sec and is easily identified and differentiated from gas in the sigmoid colon on the basis of its anatomic site and differences in proton density. Lesions of the bladder are easily seen as masses pushing into the high-T_1 area. Neoplasms of the sigmoid colon and rectum have been investigated and, in each case, the bowel has been surrounded by an irregular area of T_1 in the range of 300 to 450 msec, presumably due to reactive edema. This phenomenon serves as a useful indication of a pathologic process in the pelvis.

Investigation of the liver has shown that hepatocellular disease and cirrhosis have a characteristic appearance. Figure 6.20 shows a cirrhotic liver, demonstrating a diffuse elevation of T_1 into a range of 180 to 300 msec, associated, in this case, with enlargement of the spleen. Note in this image the large amount of fat around the upper poles of both kidneys.

The T_1 of both primary and secondary liver tumors is in the range of 300 to 450 msec. Figure 6.21 shows the characteristic appearance of numerous small metastases throughout the liver. They appear as localized areas of increased T_1 within normal liver. Note in this image the large inferior vena cava and a large cyst arising from the upper pole of the right kidney.

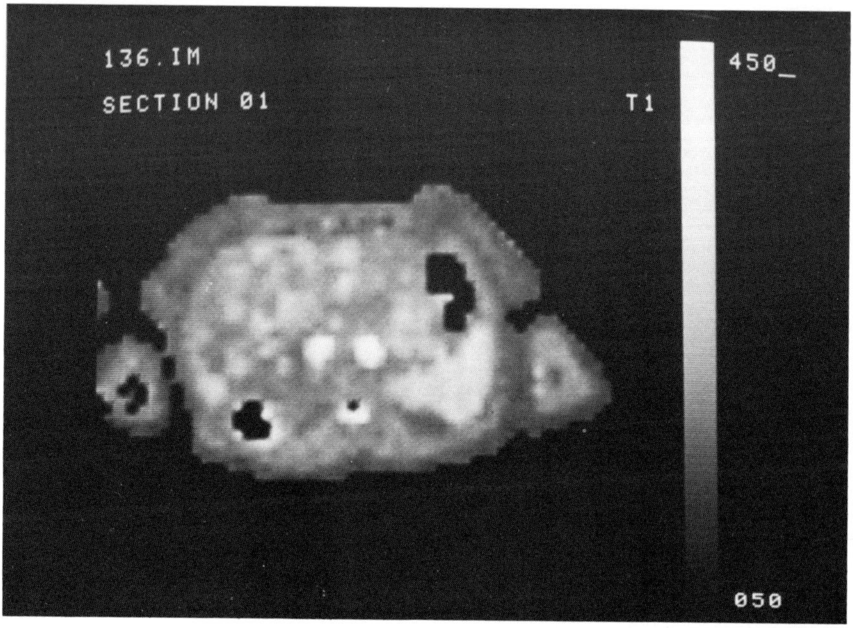

Figure 6.21. Metastatic liver disease.

Figure 6.6 shows the T_1 image of the upper abdomen in a man who experienced the sudden onset of ascites. Liver biopsy confirmed the presence of hepatoma. This section shows an irregular liver with a T_1 of 320 msec, consistent with tumors. The ascites are seen as a black area lateral to the liver and have an estimated T_1 of more than 800 msec.

Bile duct obstruction is well demonstrated in Figure 6.22, in which a large gallbladder is seen adjacent to enlarged ducts. The T_1 of normal bile is in the range of 250 to 350 msec but has been noted to rise as high as 800 msec in cases of obstruction. The cause of obstruction in this case was infiltrating carcinoma of the pancreas. In the cases of biliary calculi that we have examined, we have not been able to demonstrate either radiolucent or radio-opaque calculi.

A small benign renal cyst is seen arising from the anteromedial aspect of the right kidney in Figure 6.22. Its T_1 was greater than 1 sec, similar to that of other renal cysts examined whose T_1 range was 850 msec to more than 1 sec.

Differentiation of cysts from solid tumors of the kidney is simple because the T_1 of renal carcinomas has been shown to range from 400 to 500 msec. An example of a renal carcinoma is shown in Figure 6.23, arising

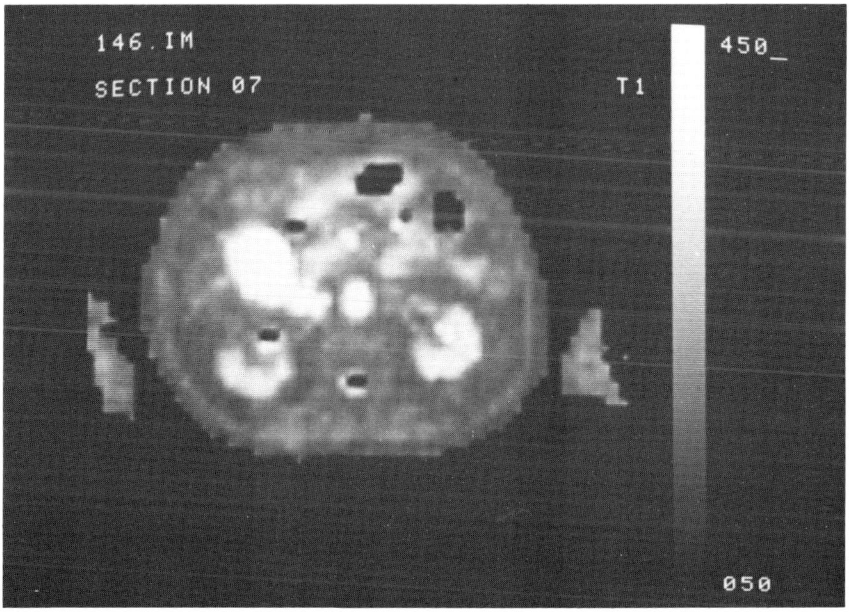

Figure 6.22. Abdominal image of biliary duct obstruction due to infiltrating carcinoma of the pancreas.

from the posterior aspect of the right kidney. In this section, the left renal artery is clearly seen arising from the aorta.

CONCLUSION

Direct comparison with conventional x-ray CT has not been possible in patients whose chests and abdomens have been examined; such a comparison would have given a direct evaluation of our NMR images. The head sections have all been compared with x-ray CT and appear as accurate anatomically as those from first- and second-generation CT scanners. In all cases of brain pathology examined by both methods, evidence of the lesion was seen on the NMR T_1 images, but the underlying tumor was often masked by surrounding edema. All the thoracic and abdominal images have been compared with those provided by conventional radiologic techniques and ultrasound and with radionuclide scans. In all cases, the NMR images have proved as accurate as those provided by these accepted techniques; in several cases, for instance, in the diagnosis of benign hepatic cysts and hemangiomas, NMR has proved more accurate.

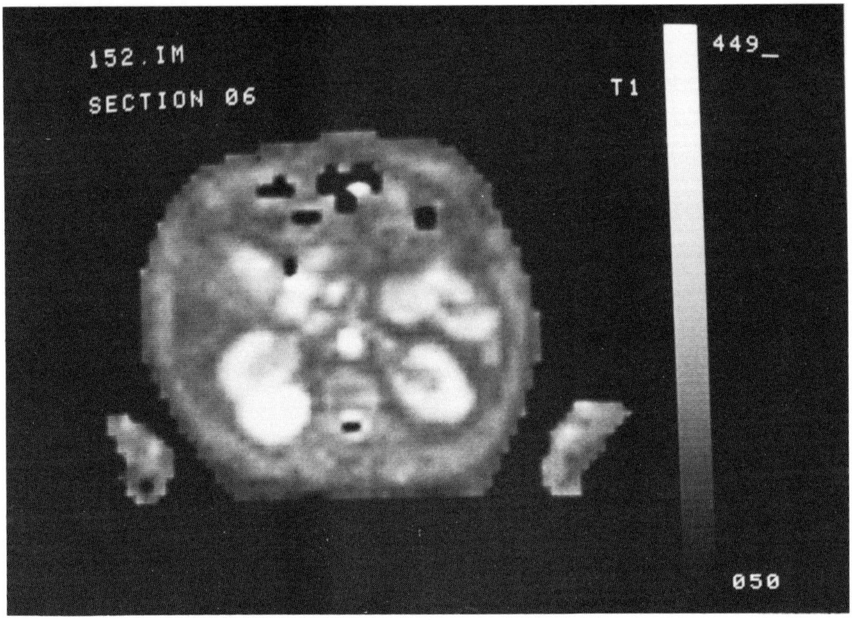

Figure 6.23. Renal carcinoma in the posterior aspect of the right kidney.

The place of NMR imaging in the diagnostic armory will probably be as a complement to existing techniques. It may be used as an effective screening procedure in many cases, especially before x-ray CT, because it is a safe, noninvasive technique that does not use ionizing radiation. To date, we have examined nearly 80 patients aged 13 to 80 years, with no ill effect. Further developments in the NMR technique should improve its spatial resolution and shorten the imaging time, thereby making it an invaluable diagnostic method.

REFERENCES

1. Hinshaw WS: Image formation by nuclear magnetic resonance: The sensitive point method. *J Appl Phys* 47:3709–3721, 1976.
2. Damadian R, Minkoff L, Goldsmith M, et al: Tumor imaging in a live animal by field-focussing NMR (FONAR). *Physiol Chem Phys* 8:61–65, 1976.
3. Mansfield P, Maudsley AA, Baines T: Fast scan proton density imaging by NMR. *J Phys E* 9:271–278, 1976.
4. Hutchison JMS, Edelstein WA, Johnson G: A whole-body NMR imaging machine. *J Phys E* 13:947–955, 1980.

5. Brooker HR, Hinshaw WS: Thin section NMR imaging. *J Magn Reson* 30:129–131, 1978.
6. Mansfield P, Maudsley AA: Planar spin imaging by NMR. *J Magn Reson* 27:101–131, 1977.
7. Lauterbur PC, Lai CM: Zeugmatography by reconstruction from projections. *IEEE Trans Nucl Sci* NS-27:1227–1231, 1980.
8. Edelstein WA, Hutchison JMS, Johnson G, et al: Spin warp NMR imaging and applications to human whole body imaging. *Phys Med Biol* 25:751–756, 1980.
9. Hutchison JMS, Sutherland RJ, Mallard JR: NMR imaging: Image recovery under magnetic fields with large non-uniformities. *J Phys E* 11:217–221, 1978.
10. Kumar A, Welti D, Ernst RR: NMR Fourier zeugmatography. *J Magn Reson* 18:69–83, 1975.
11. Hutchison JMS, Sutherland RJ, Edelstein WA, et al: 1917 Field gradient coils, UK Patent 7934864.

7

Blood Flow Measurements by NMR

Jerome R. Singer, Ph.D.

Various NMR-based body-scanning procedures for tissue imaging have been proposed and demonstrated by Andrew et al. (1), Crooks et al. (2), Damadian et al. (3), Hinshaw et al. (4), Lauterbur (5), Kumar et al. (6), Hutchison et al. (7), Mansfield et al. (8), and Moore et al. (9,10). (See the cited references as well as the study of NMR imaging by Libove and Singer [11].) Whereas these studies were concerned with NMR tissue imaging by body scan, we shall describe work on measuring the flow of body fluids.

NMR is a particularly appropriate noninvasive method for flow determinations because it is highly sensitive to the hydrogen present in blood and tissue, thus permitting the measurement of flow of blood and other body fluids. In contrast to current three-dimensional image formation by NMR techniques, one can obtain a representation of three-dimensional physiologic flow patterns by using pulsed NMR techniques as the transducer element of the system to provide optimal displays of flow within the body, thus complementing the imaging procedure.

To date, our NMR flow delineation studies have used special electromagnets designed by the author to study the flow patterns in human beings. We have obtained blood-flow distribution measurements in regional areas and have demonstrated that blood flow is readily ascertained for regions deep within the body. Our findings have applications in vascular research on the limbs of human beings and are applicable to the study of circulation in small animals. However, the most important considerations are to study blood flow and blood concentration mapping in the hearts and the brains of human beings. Our prototype system, discussed later in the chapter, is designed with these goals in mind.

In the body of this chapter, we shall describe the NMR flow measurement techniques that we have developed (12-18) over the past two decades, some of which can be employed to obtain three-dimensional displays representing the flow volume of blood or other body fluids in a region of the human body. We have chosen to give a chronologic account of our research because we felt that this approach could best provide the reader with a clear understanding of both the physical principles and practical considerations that have governed the theoretic and technologic development of this parallel application of NMR to biologic systems.

HISTORICAL BACKGROUND

Original Experiment

The possibility of a new approach to blood flow measurement first occurred to me in 1957. As with many ideas, it needed a catalyst, in this case a chance discussion with Dr. Melvin Calvin. We were both involved in research projects in nuclear and electron paramagnetic resonance and could easily discuss the possibilities of a blood flow measurement system.

NMR had only been discovered in 1946; Edward Purcell and Felix Bloch received the 1952 Nobel prize for their discovery. It was clear to both of us that here existed a method for tracking fluids within the body without making any contact to the body or disturbing the subject in any manner. Calvin enthusiastically offered to supply some of his equipment and laboratory space, allowing the first experiment to be carried out (12). The initial experiment, the measurement of blood flow in a mouse tail, was to test the validity of our belief that we could measure blood flow using the magnetic properties of the hydrogen nuclei, which are ubiquitous in blood.

The basic feasibility experiment was based on the following idea:

If we surround a flowing blood channel with two coils, as shown in Figure 7.1, we will find differences in the observed NMR signal heights when the flow velocity changes. The difference is due to the fact that with more flow, the protons just entering the NMR signal-detection area are oriented to an equilibrium value in the field, whereas the protons that remain in the NMR signal-detection area are redistributed in orientation because of the radio frequency field.

The time necessary for the protons of the bloodstream to line up or reach an equilibrium orientation in a magnetic field is approximately 0.5 sec. A more exact measurement has been described by Crooks and Singer (19).

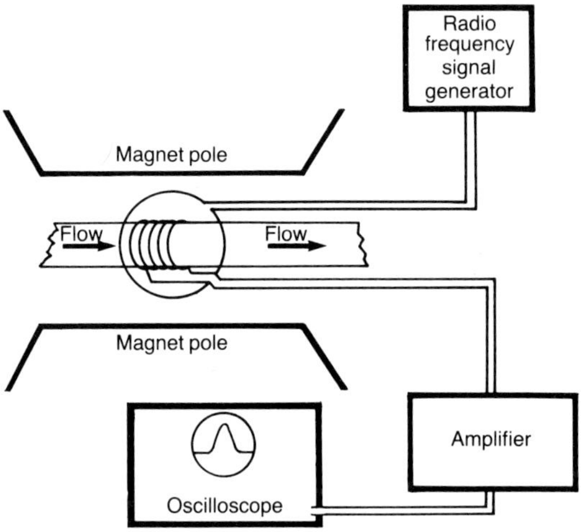

Figure 7.1. Schematic system for observing nuclear magnetic resonance. The signal is observed whether or not the fluid flows, but the signal changes with flow.

For our first experiments, we used mice because they are easily handled and their tails are convenient flow channels. The first experimental arrangement is shown in Figure 7.2. The mouse was lightly anesthetized using ether vapor, and its tail was then slipped into a test tube. The measurement procedure consisted of noting the difference between NMR signal amplitudes when flow was normal and when it was interrupted by a tourniquet around the base of the mouse's tail. A theoretically derived formula showed that the greater the blood flow, the larger the signal amplitude should be. The experimental result completely agreed with this theory, as noted in References 13 and 14.

Model Flow Measurements

This result encouraged us to proceed further. Redesigning our apparatus for more exact measurements, our next step was to try simpler geometries than exist in living tissues.

At about that time, the National Aeronautics and Space Administration was concerned about measuring fuel flow in rocket engines without introducing any obstructions in the flow line. Since such a project lent itself to our NMR measurements, we agreed to try to build a prototype. The fuel used at that time was mostly kerosene and oxygen. Kerosene is somewhat related to blood in a curious way. The protons of the kerosene had an NMR relaxation time of 0.7 sec; blood has an NMR relaxation time of

Figure 7.2. First NMR measurement of blood flow in vivo. The experiment was carried out by the authors in 1959. The mouse had been sedated by ether vapor, and the volume of blood flow was observed.

nearly the same value. Because of the similarities in relaxation times, the results of our flow experiments with jet fuel could be easily extended to the problem of blood flow measurement.

Because the area of a fuel flow pipe is constant, one need only measure the velocity of flow, multiply by the area of the pipe, and thus obtain the rate of volume flow. The prototype flow measurement used two separated coils. A pulsed radio frequency transmitter was connected to the first coil. Downstream, an NMR detection system was connected to the second coil. A solenoid created a fixed polarizing magnetic field.

The measurement procedure we used at that time is as follows: Before pulsing the transmitter on, the receiving system detects a conventional NMR absorption signal. When the transmitter is pulsed on, the group of protons directly under the transmitter coil are tipped upside down. We define the tipped group of protons as "the bolus." The bolus flows downstream and enters the receiving coil area. At that moment, the observed NMR signal reverses in amplitude because of the presence in the receiving coil area of protons that have been inverted in orientation by the pulse from the transmitter coil.

The measurement of flow velocity consists, then, of noting the time between pulsing the transmitter and receiving the reversed signal and dividing that value into the distance between the transmitter and receiver coils. The result is the average velocity of flow. The volume flow is obtained by multiplying the cross-sectional area of the flow pipe by the average velocity of flow.

By triggering the transmitter pulse "on" each time the bolus was received by the NMR detector, a numeric counter could be stepped so as directly to read the total flow volume. With the minor step of electronically dividing the total flow by the elapsed time, the volume flow rate could be read directly from a digital meter. The experimental arrangement is shown in Figure 7.3.

Blood Flow Measurements in Human Beings

Next, we again directed our efforts toward measuring blood flow in human beings. Other groups had begun similar efforts. Some measurements of blood flow in the femoral arteries of dogs were carried out in the U.S.S.R. by Zhernovoi and Latyshev (20) and at the Medical School of Wisconsin by Battocletti et al. (21).

One of our immediate difficulties in measuring human blood flow was that such subjects provide a very sensitive antenna. A coil placed close to a human subject picked up not only the NMR signal but also interference from

Figure 7.3. Experimental prototype for measurement of satellite rocket booster fuel flow. The flow measured here is approximately 20 gallons/sec!

radio broadcasting signals, 60-cycle power-line signals, and other miscellaneous signals, including electrocardiac signals. It soon appeared that conventional filtering was not sufficient to block out all the noise. However, during this period, a colleague, Dr. Melvin Klein, solved similar problems of pulling out weak NMR signals from background noise by using a small digital computer. The computer was programmed to average the random noise masking the periodic signals, thus vastly improving the periodic NMR signal relative to the noise.

The Klein technique rapidly became popular and is now a standard component of NMR systems. With the help of an averaging system, Oliver Morse and I were able to measure venous flow in the forearms of several persons, including Morse himself (16). A photograph of the measurement system is shown in Figure 7.4.

The venous flow measurement was similar to the system for measuring rocket fuel flow: The elapsed time between the tipping of the protons and

Figure 7.4. NMR measurement of the velocity of venous blood flow. *Newsweek* termed this method "an electric divining rod for blood!" Actually, the measurement is a straightforward application of NMR principles. The elapsed time for the magnetically flipped protons in the lower coil to reach the upper coil is recorded. The value gives the time for the blood to flow from coil 1 to coil 2.

the reception of tipped protons downstream is measured. When the distance between the transmitter and receiver coils is divided by the elapsed time, the average flow velocity is obtained.

We had thus reached our first objective: the accurate, reproducible measurement of blood flow speed along a specific pathway. Still, we were a long way from charting flow into a whole-body region.

The next approach was a major step forward from our previous work. Other experimenters (Hahn at Berkeley [22] and Purcell and Carr at Harvard [23]) had found that the diffusion of molecules in the resonance observation region would, logically enough, cause the phase of these molecules to get out of step with that of other molecules. The diffusion could be observed as a measured decrease in the magnetic resonance signal. Thomas Grover and I extrapolated from diffusion to flow (17). The flow of molecules will also result in phase-shift or out-of-step effects of some protons with neighboring protons. For the fluid molecules surrounded by a magnetic resonance detection coil, the flow effect could be observed as a change in the magnitude of the magnetic resonance signal.

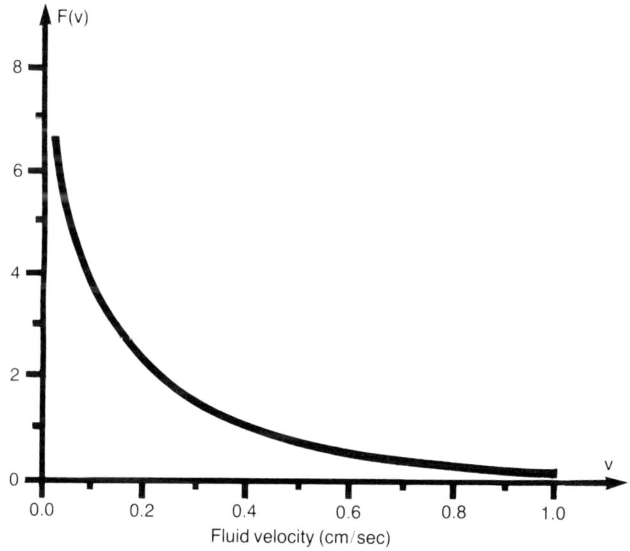

Figure 7.5. Experimentally observed flow distribution function of a human index finger. Note that most flow is at low velocities, that is, in capillary channels.

For our initial measurements, we used water flow in glass tubes. When that proved successful, rat tails were used, and when the rat-tail flow measurement was successful, measurements were carried out on human fingers. This new method of flow measurements provides a quite different result than obtained previously. The new method provides a result that we call a "flow graph." The flow graph consists of a plot of the relative number of molecules flowing in each velocity interval. An example of the output graph obtainable from the finger with the flow-graph technique is shown in Figure 7.5.

Flow Graphs

The velocity distribution function, denoted as $F(v)$, is obtained from our experiments and calculations as a function giving the number of molecules per velocity interval. By plotting $F(v)$ versus velocity, we obtain a picture of the relative number of molecules flowing at different velocities. A simple example will illustrate one such distribution. Consider the case of laminar flow in a circular tube. When we measure $F(v)$ and plot it versus velocity, Figure 7.6 is obtained. We see that the flow is normalized to unity, that equal numbers of molecules flow in each velocity interval, and

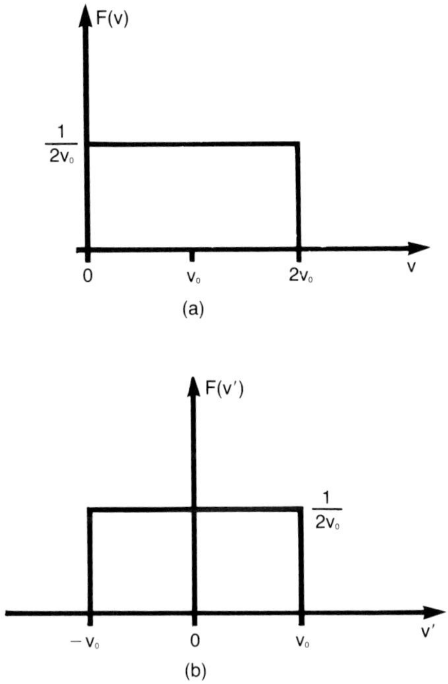

Figure 7.6. Flow distribution function, F(v), obtained for the case of laminar flow in a circular tube. Note that whereas a distribution of velocity vectors would form a paraboloid of revolution, equal quantities of fluid flow at each velocity (a). In (b), the distribution graph is displaced along the abscissa so that the plot is an even function symmetric around zero.

that the flow distribution is an even function, with half the flow greater than and half the flow less than the average velocity. Note that although Figure 7.6 does not describe the spatial distribution of flow, this result is typical of this particular mode of operation. We shall later describe the mapping of different regions of space within the body. The velocity distribution function, F(v), is useful for physiologic and biologic studies, in which velocity distributions are indicative of the nature and condition of the flow channels.

A simplified experimental setup is illustrated schematically in Figure 7.7. The protons in the flowing liquid make up the sample undergoing spin-echo magnetic resonance. The protons are subjected to radio frequency pulses of sufficient frequency, length, and intensity to precess the protons through 90 degrees and, after a delay T_1, through 180 degrees. The time between pulses (t_1) is varied systematically, and the echo amplitude and phase $2t_1$ sec after the 90-degree pulse are recorded, using a memory

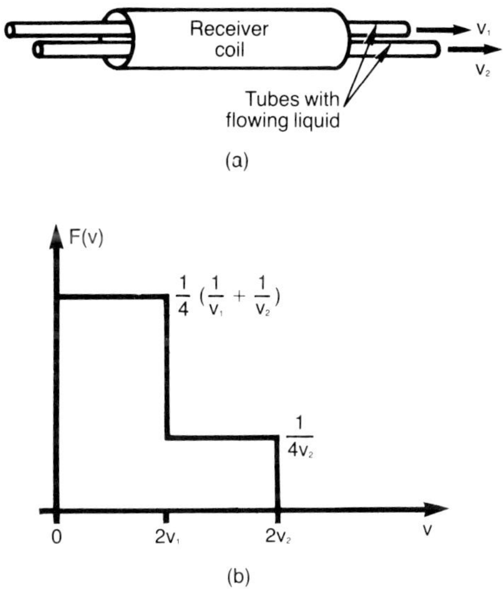

Figure 7.7. An experimental arrangement used to check out the prototype flow-measurement system (a) and resultant flow-distribution function obtained (b).

oscilloscope in our present prototype. The spin-echo background of our work is treated by Hahn (22), Purcell and Carr (23), and Farrar and Becker (24), who have written an excellent review of pulsed NMR experimental procedures.

The sampled amplitude of the echo found in this manner is defined as $A(t_1)$ and the sampled phase of the echo is defined as $\Phi(t_1)$. By taking a cosine transformation of the echo amplitude, the even part of the flow-velocity distribution function is obtained. By combining the phase and amplitude information, the total velocity distribution is obtained.

An intuitive conception of the procedure is useful. Consider the groups of spins, rotated through 90 degrees into the transverse plane. The phase distribution of the groups fans out because the flow of the fluid changes the relative phase relationships of the groups. Then, after a time t_1, a 180-degree pulse is applied. Were there no flow and no diffusion in the fluid, the 180-degree pulse would cause all the spin groups to come into phase, and all the spins would contribute to an echo pulse (except for relaxation effects, which are negligible for the times of our fluid measurements). The flow therefore appears as a change in the echo amplitude $A(t_1)$ due to phase changes in the groups of spins. In the experiment, the time between pulses (t_1) is varied, and the function $A(t_1)$ is observed as a function of the time between pulses.

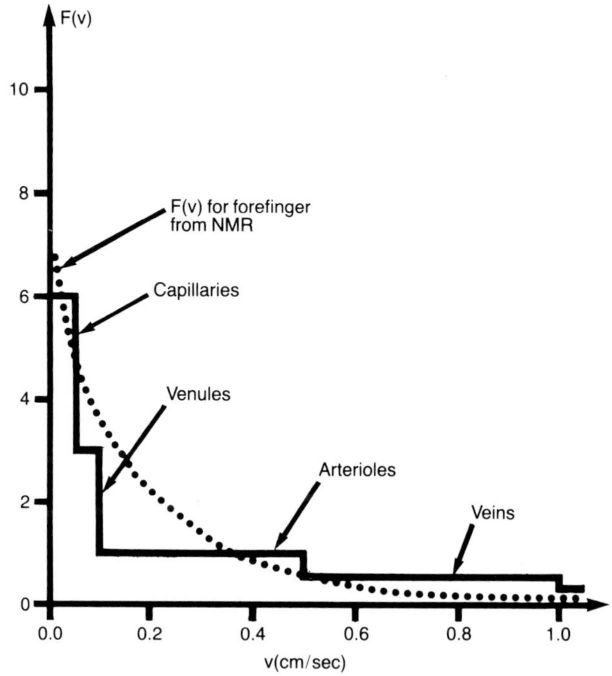

Figure 7.8. Rough comparison of the flow-distribution function obtained from a human forefinger with the relative flow area of different flow channels as estimated from data obtained from dissection of a dog.

By use of this method, several interesting flow-distribution functions are obtained. Figure 7.7 presents a velocity-distribution function for two different circular tubes, both carrying fluid in the same direction.

Of greater interest to us was the measurement of the velocity-distribution function in a human finger, shown in Figure 7.5. The data are compared to data obtained by counting blood-carrying vessels in a dog shown in Figure 7.8. Some of our data were obtained by measuring blood flow in rat tails. The velocity-distribution function for these data in comparison to the human-finger function is shown in Figure 7.9, in which the two curves are superimposed (renormalized) to compare their relative shapes.

The use of flow graphs to describe blood flow parameters provides the means for describing the total flow of blood in a region of the body, for example, limbs, heart, and brain. We can now obtain a total-flow picture of blood flow as a function of the velocity of flow and thereby can determine baseline and pathologic flow patterns.

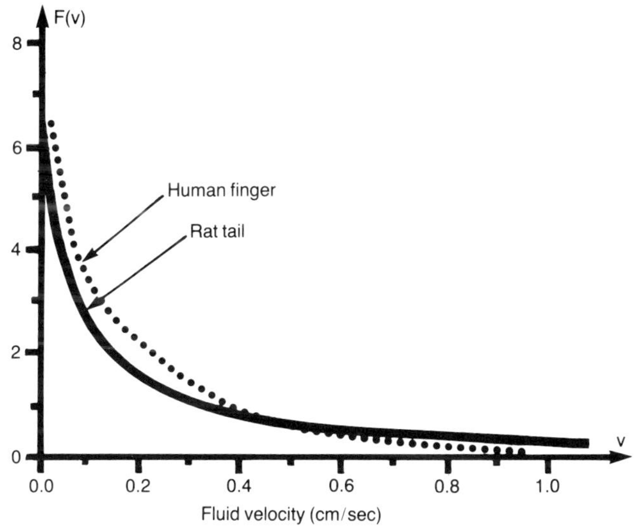

Figure 7.9. Superposition with renormalization to compare the shapes of the flow-distribution functions obtained experimentally from a human index finger and a rat tail.

In our more recent studies, we have found that our apparatus can distinguish flow direction, so that flow graphs can now be determined for both directions of flow. In essence, this procedure will provide two flow graphs for each measurement region; one graph can automatically be plotted for venous flow. We believe that this is an important step in our studies and are now approaching the stage where our system can be used in routine clinical testing.

BLOOD FLOW IMAGING AND MEASUREMENT

Blood Flow Imaging

Progress toward clinically useful blood flow displays rests on techniques for separating out NMR imaging of flow from NMR imaging of solid tissue. There are well-established principles that permit us to develop the necessary techniques. If a magnetic field gradient is along the flow path, flowing fluids are always moving into either higher or lower magnetic fields. The experimental observation of flow imaging is, therefore, obtainable by considering the time sequence of NMR images (Fig. 7.10).

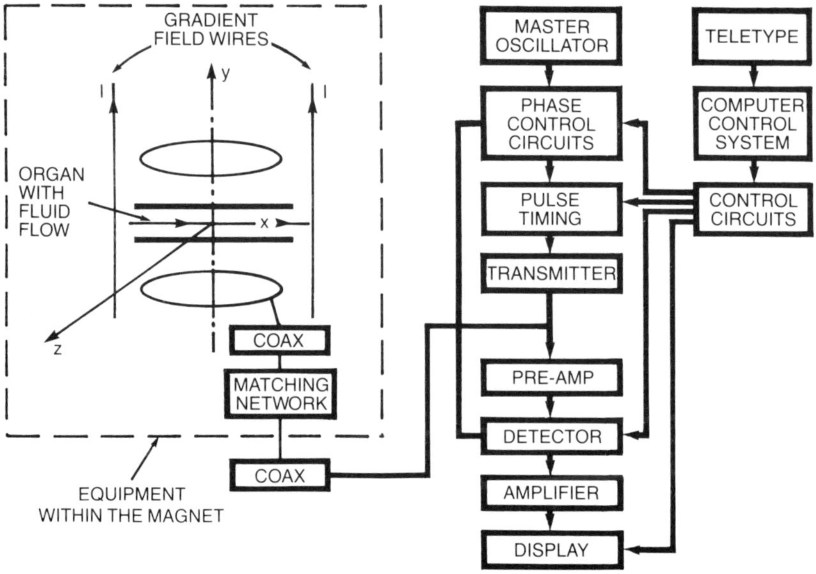

Figure 7.10. Schematic diagram of our current NMR prototype imaging system.

We have carried out experiments in which we observed both flow and nonflow images, so that the practicability of a procedure for separating out flow from solid tissue has been established. There are certain considerations that do require great care in these experiments, and some elaboration may be helpful.

If the NMR observation is directed primarily at observing solid tissue, one can still obtain blood flow observations as a consequence of the time variance of blood in magnetic field gradients. However, this observation will tend to be incidental. For example, Crooks et al. (2) carried out some NMR imaging experiments in which an animal died, due to an overdose of anesthetic, while imaging was in progress. They immediately noticed a dramatic change in the NMR image of the heart. The change was due to the blood flow stoppage, and the difference between an NMR image of still and flowing blood is, indeed, profound (see Chapter 5).

On the other hand, to obtain useful blood flow imaging measurements, one must use a different approach. There are really two procedures that should be linked to observe both flow imaging and flow measurement.

The appropriate approach is first to establish one measurement for NMR imaging. The image then establishes localization, that is, the position of major blood channels so that they may be observed. The magnetic field is then set so as to image the blood flow regions. At the same time, one sets the imaging-resolution area to be large enough to encompass the blood flow region of specific interest. Broadening the imaging-resolution region is usually accomplished by reducing the magnetic field gradients in the region, or by tailoring the radio frequency to provide a larger observation region. The observer can then measure blood flow in the region using a time sequence of observations. That is, flow measurement may be accomplished by noting and digitizing the relative signal strength from the blood flow region that is then "focused in."

Quantitatively, one desires to have an accurate measurement of the specific blood flow. There is a basis for achieving fiducial marks that allow for accurate measurements of exact, rather than relative, blood flow measurements. This approach is to carry out blood measurements using the flow-graph approach, as described above. The flow provides a plot of blood flow per velocity interval versus velocity. Therefore, a simple computer program can determine total blood flow by integrating the area under the flow graph. Blood flow graphs of specific regions in the body are a powerful means of determining both flow per velocity interval and total blood flow in that region.

There are other flow measurement procedures for obtaining information on blood flow. One will be able to observe relative flow information whenever the imaging procedure does not average out the blood flow. For example, the important early imaging work of Hinshaw et al. (4) used a signal-averaging technique that imaged flow channels, in a mode wherein the relative blood-flow velocity information was lost in favor of obtaining better image information in a shorter time. This result appears a necessary by-product of the Hinshaw sensitive-point or sensitive-line procedure, probably because signal averaging takes place over both space and time.

Qualitative Blood Flow Imaging: QRS Method

The simplest method of obtaining NMR blood-flow information is a qualitative measurement procedure; we'll call it the "QRS synchro method." If one averages NMR imaging signals in synchronism with electrocardiogram signals in a coherent manner, the blood-flow signal information can be retained. To carry out this method, the NMR imaging signals must be carefully timed relative to the electrocardiogram of the subject, separated, and collected in different sets. For example, NMR signals obtained when the QRS signal is maximum (at R) can be averaged at Q in the QRS cycle. Each image in the set represents an image for a specific point in the electrocardiographic cycle. Comparison of these two image numbers provides a relative measurement of blood flow for the specific NMR imaging region where there is blood flow. This procedure of NMR imaging in synchronism with the electrocardiogram is a relatively fast method for obtaining qualitative information on blood flow. We shall elaborate on this point.

Consider two NMR images. One, we shall assume, is taken at the point of maximal flow, which might be near the R point of the QRS cycle. Call it the R image. In other words, the R image provides a section that is consistent with the blood flow at a maximal rate. We now look at our second NMR image, which is taken in synchronism with the Q portion of the QRS signal. This image, therefore, depicts the same body region with a lower blood-flow rate. Call it the Q image. A comparison of the Q and R images will show the relative effect of flow on the NMR image. This procedure would provide a fairly rapid means of observing relative flow in a specific body area. One need not have a large number of sets; it may be sufficient to compare only a Q-synchronized image with an R-synchronized image. Along with relative blood flow, the symmetry or asymmetry of blood-flow regions can be quickly determined.

There are many applications of this modified imaging procedure. For example, carotid artery examinations are concerned with a comparison of left-right carotid flow values. Studies of upper thigh flow on a comparison basis is practicable. Atherosclerotic plaque may be localized by noting relative flow stoppage, and Crooks and Kaufman (25) have experiments of this type in progress. Furthermore, as more data are obtained, the qualitative information will become more quantitative because digitizing the data is quite straightforward.

It is clear from our preliminary studies that blood flow will show up in an important way for most, but not necessarily all, NMR imaging procedures if the imaging procedure is synchronized and classified according to different points on the patient's electrocardiogram signals. This blood flow measurement may have considerable implications for symmetry of flow studies. The QRS synchro method can be carried out with present imaging systems, at only the cost of slower running time for an image.

The more quantitative method of carrying out NMR flow-graph measurements in specific regions will give more detailed flow information at the cost of a more complex system. The flow graph has the capability of giving total-flow data, as described above. Flow graphs have the potential of completely delineating blood flow, both quantitatively and qualitatively, and we are continuing to work toward a clinically practicable system for quantitatively measuring and imaging blood flow.

For readers who are interested in a broader understanding of the problem of blood-flow measurement by NMR than a brief introductory chapter can provide, we include a short bibliography of some important literature in the field.

REFERENCES

1. Andrew ER, Bottomley PA, Hinshaw WS, et al: Multiple sensitive point method: Application to larger biological systems. *Phys Med Biol* 22:971–974, 1977.
2. Crooks LE, Hoenninger JC, Arakawa M, et al: Tomography of hydrogen with nuclear magnetic resonance. *Radiology* 136:701–706, 1980.
3. Damadian R, Goldsmith M, Minkoff L: NMR in cancer XVI: FONAR image of the live human body. *Phys Chem Phys* 9:97–100, 1977.
4. Hinshaw WS, Andrew ER, Bottomley PA, et al: Display of cross sectional anatomy by nuclear magnetic resonance imaging. *Br J Radiol* 51:273–280, 1978.
5. Lauterbur P: Medical imaging by nuclear magnetic resonance zeugmatography. *IEEE Trans Nucl Sci* NS-26:2809–2811, 1979.
6. Kumar A, Welti D, Ernst R: NMR Fourier zeugmatography. *J Magn Reson* 18:69–83, 1975.
7. Hutchison J, Edelstein W, Johnson G: A whole-body NMR imaging machine. *J Phys E* 13:903–930, 1980.
8. Mansfield P, Morris PG, Coupland RD, et al: Carcinoma of the breast imaged by NMR. *Br J Radiol* 52:242–243, 1979.
9. Holland GN, Hawkes RC, Moore WS: Nuclear magnetic resonance tomography of the brain: Coronal and sagittal sections. *J Comput Assist Tomogr* 4:429–433, 1980.
10. Hawkes R, Holland G, Moore W, et al: Nuclear magnetic resonance (NMR) tomography of the brain: A preliminary clinical assessment with demonstration of pathology. *J Comput Assist Tomogr* 4:577–586, 1980.
11. Libove J, Singer JR: Resolution and signal-to-noise relationships and NMR imaging in the human body. *J Phys E* 13:38–44, 1980.
12. Singer JR: Application of magnetic resonance phenomena to biological problems including measurement of blood flow. Paper presented at IRE Twelfth Annual Conference on Electrical Techniques in Medicine and Biology, Philadelphia, November 3, 1959.
13. Singer JR: Blood flow rates by NMR measurements. *Science* 130:1652–1653, 1959.
14. Singer JR: NMR techniques for blood measurements. *J Appl Phys Suppl* 31:406s–407s, 1960.
15. Singer JR: Flow rates by nuclear and electron paramagnetic resonance methods. *J Appl Phys* 31:125–128, 1960.
16. Morse O, Singer JR: Blood velocity measurements in intact subjects. *Science* 170:440–442, 1970.
17. Singer JR, Grover T: Recent measurements of flow using NMR techniques, in C. G. Clayton (ed.): *Modern Developments in Flow Measurement*. London, Peter Peregrinus, 1972, pp 38–44.
18. Singer JR: NMR diffusion and flow measurements and an introduction to spin phase graphing. *J Phys E* 11:281–291, 1978.

19. Singer JR, Crooks LE: Some magnetic studies of normal and leukemic blood. *J Clin Eng* 3:237–243, 1978.
20. Zhernovoi AI, Latyshev LF: *Nuclear Magnetic Resonance in a Flowing Liquid*. New York, Consultants Bureau, 1965.
21. Battocletti JH, Halbach RE, Salles-Cunha SX, et al: Clinical application of a nuclear magnetic resonance (NMR) limb blood flow meter. *Proc IEEE* 67:1359–1362, 1979.
22. Hahn EL: Spin echoes. *Phys Rev* 80:580–594, 1950.
23. Carr HY, Purcell EM: Effects of diffusion on free precession in nuclear magnetic resonance experiments. *Phys Rev* 94:630–635, 1954.
24. Farrar TC, Becker ED: *Pulse and Fourier Transform NMR*. New York, Academic Press, 1971.
25. Crooks LE, Kaufman L: Personal communication.

BIBLIOGRAPHY

Battocletti JH, Linehan JH, Larsen SJ, et al: Analysis of a nuclear magnetic resonance blood flowmeter for pulsatile flow. *IEEE Trans Biomed Eng* 19:403–407, 1972.

Bowman RL, Kudravcev V: Blood flowmeter utilizing nuclear magnetic resonance. *IRE Trans Med Electron* 6:267–269, 1959.

Garroway AN: Velocity measurements in flowing fluids by NMR. *J Phys D* 7:L159–L163, 1974.

Hayward RJ, Packer KJ, Tomlinson DJ: Pulsed field-gradient spin echo NMR studies of flow in fluids. *Mol Phys* 23:1083–1102, 1972.

Hemminga MA, Jager DA, Sonneveld A: The study of flow by pulsed nuclear magnetic resonance. I. Measurement of flow rates in the presence of a stationary phase using a difference method. *J Magn Reson* 27:359–370, 1977.

Hirschel LR, Libelo LF: Nuclear magnetic resonance in flowing fluids. *J Appl Phys* 32:1404–1405, 1961.

Hirschel LR, Libelo LF: NMR signal dependence on fluid velocity. *J Appl Phys* 33:1895–1896, 1962.

Packer KJ: The study of slow coherent molecular motion by pulsed nuclear magnetic resonance. *Mol Phys* 17:355–368, 1969.

Singer JR: See Chapter References 11–19.

Stejskal EE, Tanner JE: Spin diffusion measurements: Spin echoes in the presence of a time-dependent field gradient. *J Chem Phys* 42:288–292, 1965.

Zhernovoi AI, Latyshev LF: *Nuclear Magnetic Resonance in a Flowing Liquid*. New York, Consultants Bureau, 1965.

III

Elements Other Than Hydrogen

8

In Vivo Chemistry with NMR

Derek Shaw, Ph.D.

Previous chapters in this book have dealt with the methods by which maps of proton density can be built up as well as their applications in medicine. Producing density maps makes use of only the most basic part of the NMR phenomenon, the ability to detect and quantify the presence of specific nuclei. In 1950, it was found that not all nuclei of the same type came into resonance at the same field/frequency ratio. Nuclei in different chemical environments had different resonance conditions. The effect was small, a few parts per million, but significant. Because this effect depended on the sample's chemical composition, it was, not surprisingly, called the chemical shift. With this discovery, whole new areas of unexpected applications of NMR were opened up. A period of careful research, over the next ten years or so, led to the basic understanding of the variables that affect the chemical shift, and from about 1960 onward, the use of NMR in chemistry, especially structural organic chemistry, grew almost explosively.

The aim of this chapter is to introduce readers to the enormous potential of applying (to an in vivo situation) the power of NMR to study individual chemical species. To date, our knowledge of biochemical reactions has been obtained either by studying what are essentially model systems in vitro and inferring the real situation or by "sampling" the real situation and hoping the sampling does not perturb the system. With modern equipment and techniques, it is now possible to use NMR to study a specific organ or part of the body of an animal or human being in a totally noninvasive, nondestructive, and harmless (see Chapter 10) manner. In other words, we can now study in vivo chemistry. Whether this ability will have the same impact on medicine as conventional NMR did on chemistry remains to be seen. The evidence so far suggests that it will.

Having stated the aims of this chapter, the limitations must also be defined. Space precludes detailed coverage of the background work done to date and thorough coverage of the instrumental techniques. A general introduction will be given to these areas; readers, however, should refer to the many available books (1-5) and review articles (6-12) for in-depth treatment. This chapter will concentrate on the areas that are relevant to the study of in vivo chemistry from a medical point of view.

BASIC HIGH-RESOLUTION NMR

Structure of High-Resolution Spectra

Chapter 2 introduced the basic physics of NMR. To consider the application of NMR to studying in vivo biochemistry, the nature of the NMR spectrum must be studied further. When the correct field/frequency ratio is achieved, a resonance absorption of energy occurs that gives rise to a Lorentzian absorption line. The area of the line is proportional to the concentration of the nucleus at resonance. It is the intensity of this resonance as a function of spatial position that forms an NMR image.

The NMR resonance condition is described by the Larmor equation:

$$\nu = \gamma \mathbf{H}_0, \qquad [1]$$

where the constant γ is the magnetogyric ratio of the nucleus being observed. It was the measurement of this fundamental nuclear constant that was the objective of the first NMR experiment by Bloch and Purcell.

To improve the accuracy of their experiment, early workers improved the homogeneity (uniformity) of their magnets and found, to their surprise, that they obtained different values of γ, that is, separate resonance conditions for different chemical environments. The first reported example of this effect was for ^{14}N, where two lines were found from a solution of ammonium nitrate (NH_4NO_3), one from each nitrogen atom. Because the magnitude of the chemical shift is small, a few parts per million, it is noticed only when the homogeneity of the magnet exceeds 1 part in 10^6. NMR imaging magnets have homogeneities of a few parts in 10^4 or 10^5; therefore, they do not resolve the small chemical-shift effects. The chemical shift is illustrated in Figure 8.1 for the protons in methanol.

It is outside the scope of this chapter to provide a detailed discussion of the nature of the chemical shift; readers are referred to a standard textbook on high-resolution NMR for this information (1,3). In simple terms, the effect arises because the bonding electrons screen the nucleus from the applied field by various amounts, depending on the type of chemical bond involved.

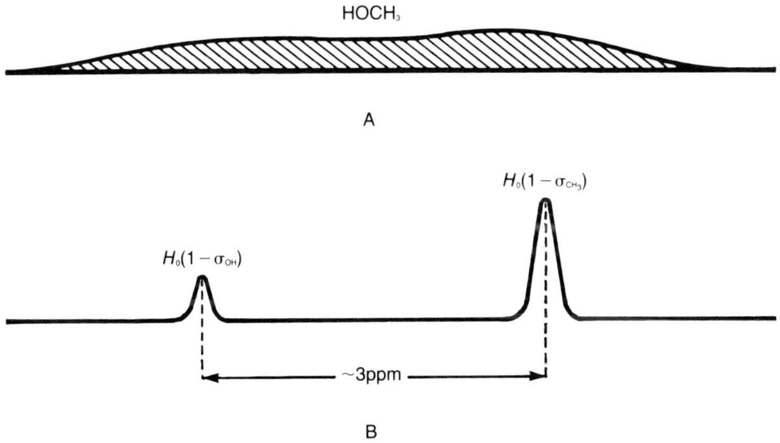

Figure 8.1. Proton NMR spectrum of methanol (CH_3OH). Under conditions of low resolution (i.e., with H_0 uniformity less than 10^5), the spectrum is essentially a single broad line (A). By contrast, under high-resolution conditions (i.e., with H_0 uniformity greater than 10^7), two chemically shifted lines are resolved arising from the methyl and hydroxyl protons (B).

The Larmor equation now becomes:

$$\nu_A = \gamma \mathbf{H}_0 (1 - \sigma_A), \quad [2]$$

where σ_A is called the screening constant of nucleus A. For example, methyl protons of methanol are more shielded from the basic magnetic field than is the hydroxyl proton; that is, $\sigma_{CH_3} > \sigma_{OH}$. Hence, resonance of the methyl protons occurs at a lower frequency than that of the hydroxyl proton at constant field or, equivalently, at higher field if the frequency is constant. The methyl protons are said to have a shift to high field of the hydroxyl proton.

In NMR, it is difficult to measure absolute chemical shifts because the only method of measuring magnetic fields with sufficient accuracy is by NMR, and a circular argument results. This problem is overcome by measuring all shifts with respect to an arbitrary reference compound, for example, the protons in tetramethylsilane in conventional NMR spectroscopy, or the phosphorus signal from phosphocreatine in the work to be considered in this chapter.

As can be seen from the above equation, the absolute magnitude of the chemical shift is field dependent. To overcome this complication, chemical shifts are reported in dimensionless units, parts per million (ppm) defined thus:

$$\text{Chemical shift of A } \delta_A = \frac{\nu_A - \nu_{ref}}{\nu_{ref}} \times 10^6 \text{ ppm}, \qquad [3]$$

where ν_A is the resonance frequency of nucleus **A** in field $\mathbf{H}_o$, and ν_{ref} is the resonance frequency of the reference compound in the same field $\mathbf{H}_o$.

In the situation where a nucleus, or the molecule to which it belongs, is involved in chemical exchange, the observed chemical shift reflects this exchange. If the exchange rate is slow compared with the reciprocal of the chemical-shift difference (expressed in Hertz), two lines are seen corresponding to the two sites; for example, when the phosphorus of phosphocreatine (PCr) is in slow exchange with the γ-phosphate of adenosine triphosphate (ATP), two signals are seen. If, however, the exchange is fast, only one signal, at the population-weighted mean, is observed. Intermediate situations are complex, because of the broad lines in the spectra, but are well understood (1).

One important case of exchange is the protonation of inorganic phosphate, that is,

$$H^+ + HPO_4^{2-} \rightleftharpoons H_2PO_4 \quad pK_a = 6.88, \qquad [4]$$

which has a pK_a in the physiologic region. There is a difference of chemical shift of 2.3 ppm between the phosphorus in the two ions, and, as a result, the observed chemical shift of inorganic phosphate (P_i) with respect to PCr is a function of pH, as shown in Figure 8.2 (13-15). Measurement of δP_i, therefore, provides a unique nonperturbing pH meter.

If NMR spectra are recorded in an even more homogeneous field, better than about 1 part in 10^8, a further interaction called the "coupling constant" is observed. This small interaction, which is given the symbol *J* and measured in Hertz, is field independent and arises from the transmission of information about the spin state of one nucleus to another via the bonding electrons. Spin-spin coupling is short range, extending over only two or three bonds, and is illustrated for the case of ATP in Figure 8.3. The α- and γ-phosphorus are each split into doublets because they sense the two possible spin states of the β-phosphorus (parallel and antiparallel to the field), whereas the β-phosphorus is split into two doublets due to both spins of both the α- and γ-phosphorus. When studying in vivo NMR, it is unusual, although not impossible, to achieve line widths that, due to the heterogeneous nature of the sample, are sufficiently narrow to permit the observation of spin-coupling effects. For further discussion of this aspect of NMR spectra, which is of great importance in chemical applications, see References 1 and 3.

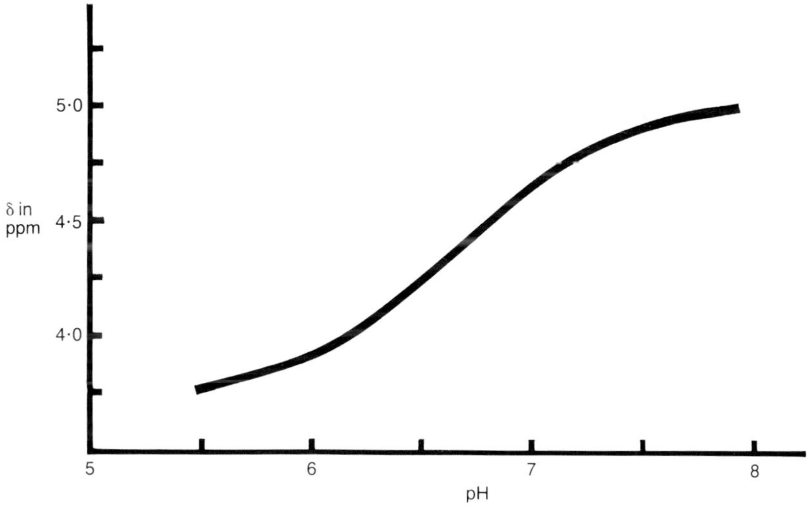

Figure 8.2. Effect of changing pH and the chemical shift of P_i. The shifts were measured with respect to PCr.

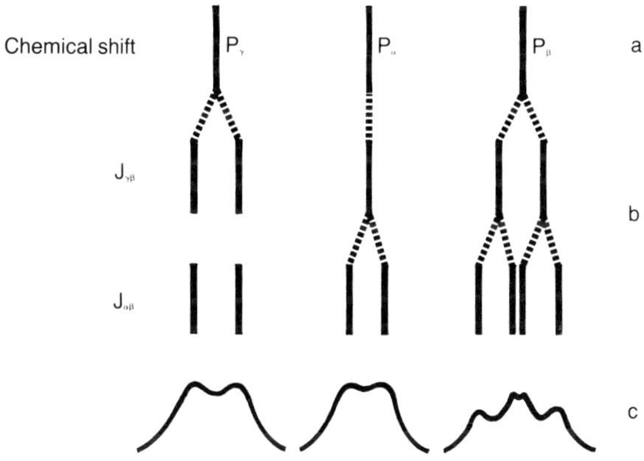

Figure 8.3. [31]P spectrum of ATP shows (a) the effect of chemical shifts separating the three phosphorus environments, (b) including the splittings due to homonuclear spin-spin coupling between adjacent phosphorus nuclei (note: $J_{\gamma\beta} > J_{\alpha\beta}$ and the long-range $J_{\alpha\gamma}$ is not resolved), and (c) the appearance of the spectrum when finite line widths are included.

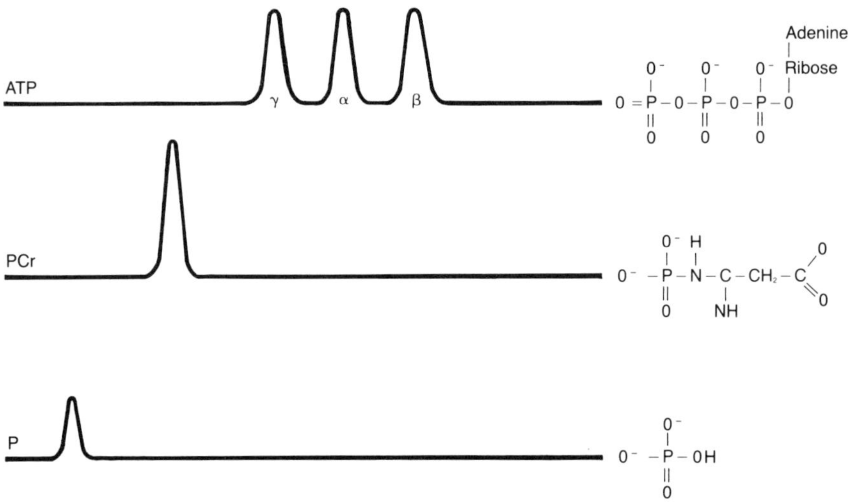

Figure 8.4. Schematic representation of the ^{31}P spectrum from animal tissue. The three species detected are ATP, PCr, and P_i. The relative areas of the peaks vary from tissue to tissue, and additional peaks are often found.

Each individual line in a high-resolution spectrum comes from the nuclei in a specific chemical environment, or one so closely related that any difference cannot be resolved. Because the area under a peak is proportional to the number of nuclei involved, it is possible to use peak areas to quantify the ratios in various chemical species present. One can thus follow the change in concentration of one molecular species, for example, PCr during ischemia, by measuring the area under the appropriate resonance line in the ^{31}P NMR spectrum. Absolute quantitative accuracy is not one of NMR's strong points due to many instrumental problems; however, relative quantification is satisfactory, provided that care is taken to avoid saturation (3).

High-resolution spectra are obtained only from molecules that are in rapid molecular motion, that is, those in liquid or gaseous phases. Generally speaking, molecules in the solid phase have too slow or limited molecular motion to average out dipole-dipole interactions, hence giving broad lines. When examining spectra from animals, therefore, only a limited range of molecules is observed. In proton spectra, signals are seen, for example, from water and fats, and for ^{31}P spectra, resonances arc observed from phosphate metabolites in the cytoplasm. Signals are not seen from phospholipid molecules or from smaller molecules tightly bound to them, nor are signals seen from bone or other solid substances. The major lines seen in muscle spectra are shown in Figure 8.4.

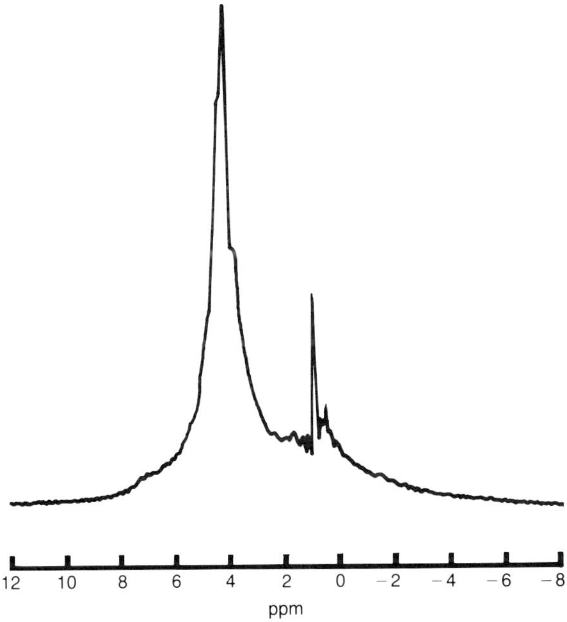

Figure 8.5. Proton spectrum of a human arm at 80.3 MHz. Note the resonances to high field of the dominant water peak arising from the aliphatic protons of the lipid droplets in the fat cells, with the olefinic resonances to low field.

Before proceeding further, it is appropriate to consider what nuclei are available and practical for in vivo chemical studies. From a biochemical point of view, they must be from atoms well distributed in molecules of interest. Such a consideration produces hydrogen, carbon, nitrogen, oxygen, phosphorus, sodium, and so on. The three nuclei that represent the best compromise between NMR sensitivity and biochemical interest and abundance are ^{1}H, ^{13}C, and ^{31}P. The proton has a major drawback for in vivo studies: Its spectra are extremely complex due to the many homonuclear spin-spin couplings present. As a result, with the homogeneity achievable for in vivo work, that is, where the sample itself is not homogeneous, only relatively gross features, such as water and fatty acids, are resolvable, as is shown in Figure 8.5. ^{13}C spectra also contain many lines but in complex molecules are often simpler to assign and apply to biochemical problems than are ^{1}H spectra (16). The major problem with ^{13}C NMR is sensitivity due to the low sensitivity of the ^{13}C nucleus (1% that of the proton) and its low natural abundance. However, as can be seen in Figure 8.6, ^{13}C spectra of a human arm can be produced in a few minutes, despite these limitations. In practice, the large biochemical concentration of ^{13}C compared with ^{31}P make them of comparable sensitivity

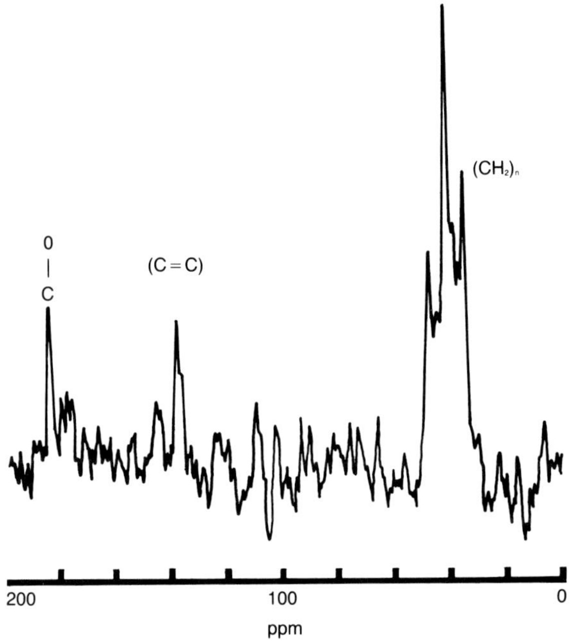

Figure 8.6. ^{13}C NMR spectrum of a human arm. The spectrum was recorded without proton decoupling in 8 min and resolves the lipid carbonyl, olefinic, and aliphatic carbons. (This spectrum was recorded on a TMR 32 spectrometer at 20.3 MHz in collaboration with R.G. Shulman, Univeristy of Yale.)

for biologic samples. Further improvements in sensitivity and spectra simplicity can be expected when the technique of proton-noise decoupling (3,8) is added. Decoupling improves the sensitivity of the experiment by means of the nuclear Overhauser effect and simplifies the spectra, reducing the signal from each chemically unique carbon to a single line by removing proton-carbon spin coupling. The low natural abundance of ^{13}C can be turned into an advantage by using enriched metabolites as a probe for biochemical pathways (see the section on Muscular Dystrophy). As is discussed later, ^{13}C NMR is a powerful biochemical method.

Phosphorus does not suffer from most of the complications discussed so far. It does occur in biochemically interesting places, for example, the ATP cycle, and has a spin-½ nucleus with a high natural abundance (^{31}P is 100% abundant). Because the phosphate signals that are normally studied do not have any directly bonded protons, no heteronuclear coupling complicates the spectra; homonuclear ^{31}P-^{31}P coupling (see Fig. 8.3) is small due to the intervening oxygen, usually less than the natural line width.

The result is that ^{31}P spectra are basically simple to interpret, yet informative. It is therefore not surprising that most of the pioneer work of in vivo chemistry was carried out using ^{31}P as the probe.

Instrumentation

General Considerations

In any physical technique, it is always advantageous to have the highest sensitivity (signal-to-noise ratio) practicable to improve the quality of the results. Achieving adequate signal-to-noise ratios is a problem in NMR, in particular, because it has an inherently low sensitivity. This low sensitivity arises from the weak nature of the interaction between the magnetic nuclei and the applied field. Because the interaction is weak, the difference in energy between the ground and excited states is small, and so, consequently, is the difference in population between these levels. As in all spectroscopic techniques, sensitivity is proportional to this energy difference. Therefore, working at higher fields produces better sensitivity as well as increasing spectra dispersion. Reference 17 gives a detailed account of the variables that influence the sensitivity of a high-resolution spectrometer when studying animal samples.

Using currently available magnetic fields, that is, up to 1.8 Tesla for whole-body work, adequate signal to noise can be achieved in ^{31}P NMR to study metabolites present in the sample at concentrations in the millimolar (or greater) range in a few minutes. A similar period is necessary to produce ^{13}C spectra, whereas proton spectra can be produced in a much shorter time.

The instrumentation required to produce high-resolution spectra of living samples can, as is the case with an NMR imaging system, be divided into three main areas: the magnet, the radio frequency probe, and the radio frequency electronics, computer, and so on (Fig. 8.7). A complete spectrometer suitable for small-animal and limb research is shown in Figure 8.8.

The key component is the magnet. This component must provide a field that is homogeneous enough for high-resolution work, that is, 1 part in 10^7 or better, over the area of interest, and it should have the highest possible field. The magnets used for high-resolution work are superconducting solenoids similar to those discussed in Chapter 4. The differences are the higher field (1.8 Tesla) and the adjustable correction (or shim) coils that permit the field to be adjusted to give a field uniform to 1 in 10^7 over a 100-mm diameter volume. Specially designed magnets can be fitted with a set of profile coils that permit the diameter of the high-homogeneity volume, that is, the sensitive volume, to be adjusted (18). Magnets currently

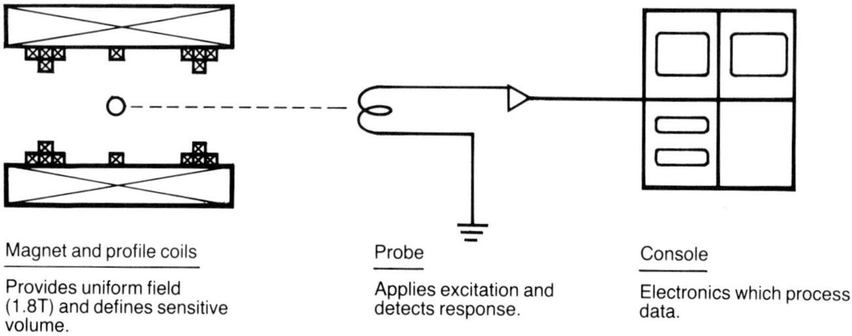

Figure 8.7. Block diagram of a TMR spectrometer.

Figure 8.8. Oxford Research Systems TMR 32 spectrometer with a 200-mm bore magnet for in vivo phosphorus and proton studies.

available have clear bores of 200 and 600 mm for animal and human studies, respectively.

The function of the probe is to apply the exciting magnetic field ($\mathbf{H}_1$) and detect its response. The probe can take two basic forms: It can either have coils that contain the whole sample and produce a uniform field, $\mathbf{H}_1$, as in the cases of conventional NMR spectroscopy and NMR imaging, or it can have smaller, flat-surface coils that are designed to study just a small part of the sample (19).

The third part of the spectrometer is the electronics necessary to produce, detect, and amplify the radio frequency signals and then digitize and process them into the final spectrum. This part of the system is identical to those used in conventional high-resolution NMR (see Chapter 6 of Reference 3 for further details).

Instrumentation for Production of Spatially Localized High-Resolution Spectra

To extend high-resolution NMR to the study of live animals and human beings, as opposed to living cells and organs, it is necessary to have a method of obtaining a high-resolution spectrum from part of the sample. Ideally, one would like to be able to use imaging processes to either provide a map (image) for each peak in the spectrum, and hence each metabolite, or to localize a volume and produce a high-resolution spectrum from that point.

In view of the small chemical-shift range and the complexity of proton spectra, it is almost certainly impossible to achieve the desired objective with this nucleus. With ^{31}P, both of the above procedures have been shown to be possible. Lauterbur et al. have used zeugmatography techniques to produce separate images representing the distribution of ATP, PCr, and inorganic phosphate within a phantom (20). However, the sensitivity presently achievable for ^{31}P makes imaging with that nucleus impractical at the present (Lauterbur used 100 times the physiologic concentrations and took several minutes). Bottomley has applied the sensitive-point imaging method to the problem of localized spectra (21). The limitations of this method are that it has conflicting requirements: Oscillation gradients are necessary to define the sensitive point, whereas, within the sensitive volume, a uniform, time-independent field is required. Compromise conditions can be achieved, but only for very small volumes and with limited line width, both of which make the approach of very low sensitivity.

Two complimentary techniques are available and in commercial production that allow spatially localized, high-resolution spectra to be obtained. These techniques are the use of surface coils (19) and topical

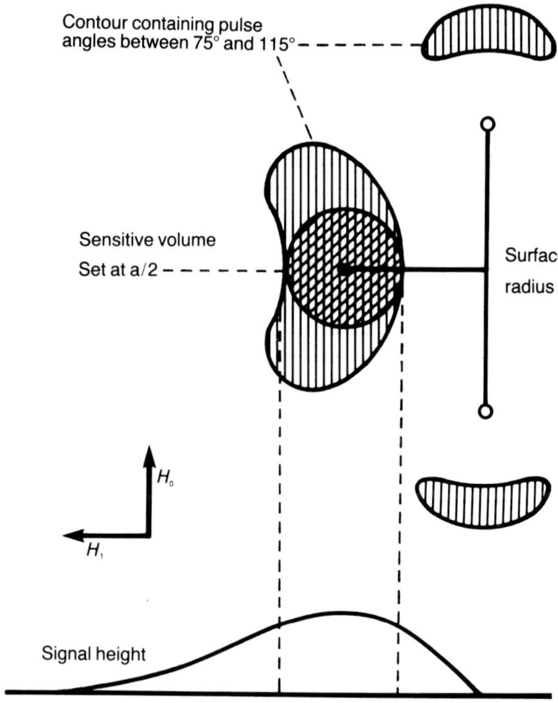

Figure 8.9. Surface coils produce a nonuniform exciting field and, consequently (because the efficiency of detection and excitation are proportional [17]), detect signals preferentially from specific areas. The dotted area within the contours indicated the area of maximum signal detected by a surface coil. When field profiling (TMR) is added (heavy shading), the spatial localization of the detected signal is further improved.

magnetic resonance (TMR) (18). Both methods shape the resonant-volume, surface-coils shape by using the exciting field and TMR by application of the static field. With these techniques, it is possible to obtain high-quality spectra from known localized volumes within an animal.

In conventional NMR spectroscopy and NMR imaging, the probe coil is designed in such a way that it contains the sample and provides a uniform exciting ($\mathbf{H}_1$) field and, consequently, receives a signal nonselectively from the sample. It was pointed out by Ackerman et al. (19) that if a flat or surface coil is used instead, it will, under normal circumstances, excite and detect signals from an approximately disk-shaped region in front of the coil. The thickness and radius of this disk are approximately equal to the radius of the coil, as is shown in Figure 8.9. The spatial specificity of a surface coil is shown by spectra in Figure 8.10, taken with the coil placed

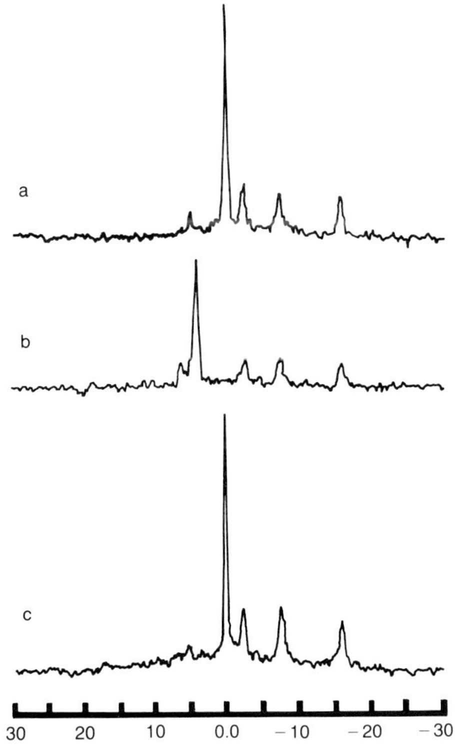

Figure 8.10. ^{31}P NMR spectra of the leg muscle of an intact rat obtained with a surface coil. Each spectrum represents 128 scans over 25 min. A 15-Hz line-broadening exponential multiplication was used to enhance signal to noise. (a) Nonischemic muscle below the knee joint, (b) ischemic muscle from the same areas as in (a) after application of a tourniquet just above the knee, (c) spectrum of muscle above the tourniquet. (19) Reprinted by permission from *Nature* 283:167–170, © 1981 Macmillan Journals Limited.

above and below a tourniquet applied to the leg of a rat.

The surface coil is ideally suited to the study of selected regions near the surface (as is illustrated by the studies described in the section on Myocardial Pharmacology). However, the surface coil has an obvious limitation when applied to regions deep within the sample; that is, the width of the sensitive volume increases with penetration depth. This drawback can be overcome by combining it with the TMR technique (see Fig. 8.9).

The technique of TMR involves shaping of the basic static magnetic field used for the measurement and is illustrated in Figure 8.11 (13). The

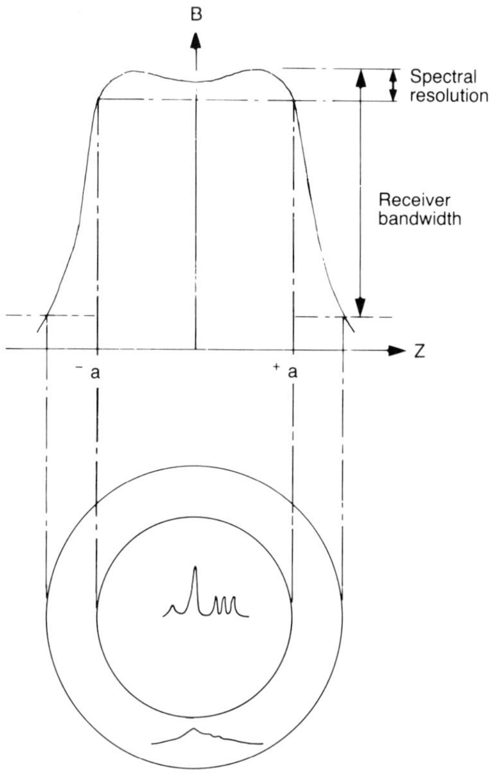

Figure 8.11. Profile of the magnetic field along the z axis in TMR. The inhomogeneity within the central region defines the spectral resolution. As shown in the projection below, the less homogeneous penumbral region gives rise to broadened lines that are mathematically removed.

horizontal-bore, superconducting magnet produces a field that, with the aid of shim coils, has the homogeneity required for high-resolution spectroscopy. Inside the magnet bore is an additional set of profiling coils that modify the field, so that within the sensitive volume, it is uniform and outside this volume it changes rapidly. The resultant spectrum consists of two components, the high-resolution spectrum from within the sensitive volume and a broad component "hump" from the "penumbra." The latter can be easily removed using the process of convolution differencing (22). This process involves multiplying a copy of the raw free-induction decay by a rapidly decaying exponential function and subtracting it from the original. The difference is then Fourier transformed in the normal way. Other mathematical manipulations, such as Lorentzian to Gaussian transformations and statistical correlations, can also be used. The diameter of the sensitive volume can be varied to home in on a specific region; however, its position stays fixed at the center of the magnet.

Figure 8.12 shows that the spectrum from liver alone, as confirmed by the lack of PCr, can be obtained from the abdomen of a healthy rat. A coil, producing a uniform $\mathbf{H}_1$ field, was used; the localization was achieved purely by shaping the $\mathbf{H}_0$ field. TMR was designed primarily to produce localized, high-resolution spectra and hence to use the highest field practical to aid sensitivity and to provide maximum separation of resonant lines (see the sections on General Considerations and Structure of High-Resolution Spectra, respectively). The sensitivity is especially advantageous for ^{31}P studies.

Operating at high fields means, by definition, operating at high frequency, which, in turn, presents problems in probe design. It is impossible to tune a conventional Helmholtz or solenoid coil large enough to accommodate a human torso at 32 MHz, let alone 80 MHz (the frequencies corresponding to the ^{31}P and ^{1}H frequencies in the 1.8-Tesla magnets currently used). To overcome this complication, the complementary nature of surface coils and TMR is used. Surface coils of the required diameter to study major organs can easily be made to tune at the required frequency. If a surface coil large enough to study a deep organ, for example, the heart, were used alone, a signal from the intervening muscle wall would also be seen (as can be seen diagrammatically in Fig. 8.9 and, in practice, in Fig. 8.13). If, however, the magnetic field is profiled so that it is homogeneous only over the central volume of the surface coil's active region, a highly selective spectrum is obtained. Unlike the heart spectra used for previous studies, the spectrum in Figure 8.13 is of a healthy rat unprepared in any way, except that it was anesthetized to keep it still. The spectrum took only 2 min to obtain, despite its small size, indicating that studies of the much larger human heart are possible.

HIGH-RESOLUTION NMR OF LIVING SYSTEMS

If one wishes to study the chemistry of living systems, it is necessary to study high-resolution NMR spectra, which, as their name implies, are obtained under conditions of high resolution, that is, in a very homogeneous field and, ideally, from a pure, homogeneous sample. Considering the complex, inhomogeneous nature of living systems and the size of the magnet required to accommodate a living animal of any real interest, it is not surprising that high-resolution studies of living systems were slow to

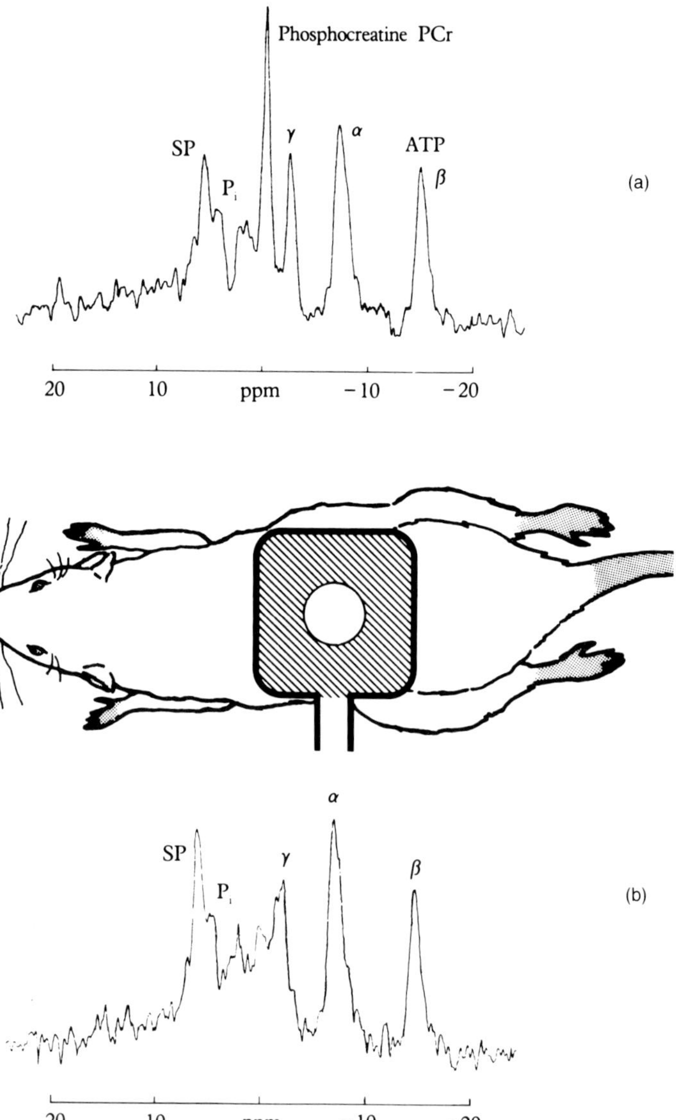

Figure 8.12. (a) The in-vivo TMR spectrum of ^{31}P obtained from a rat abdomen using a sensitive volume diameter 2a = 5 cm. (b) The in-vivo TMR spectrum of ^{31}P obtained from a rat abdomen using a sensitive volume diameter 2a = 2 cm.

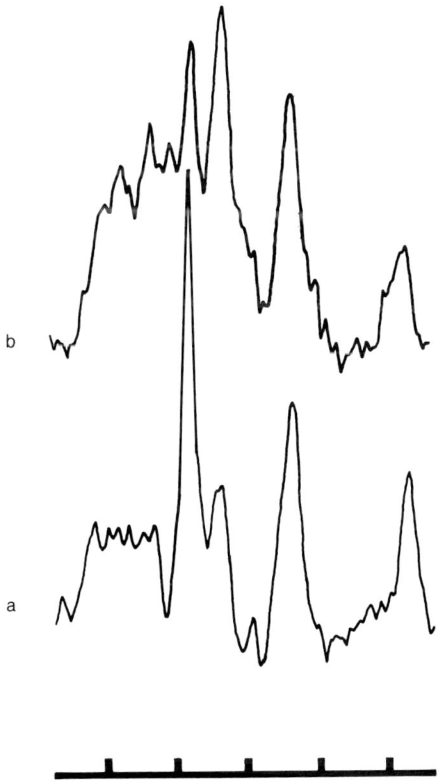

Figure 8.13. ^{31}P spectra of an anesthetized rat. Spectrum (a) is taken with a 4-cm surface coil placed on the chest and a 4-cm sensitive volume. The spectrum gives the metabolite levels for the whole chest region. Spectrum (b) is obtained with the sensitive volume reduced to 2 cm and originates from the region of the animal's heart. Note the power ratio of PCr to ATP compared with (a), which is typical of heart (38) as compared with skeletal muscle (19), and the "hump," which has often been found in spectra of perfused hearts (39).

take off. This section gives a basically historical account of the in vitro work that has led to our present understanding of the potential of NMR to study in vivo chemistry before a discussion of present research.

In 1973, Moon and Richards looked at the ^{31}P spectrum of blood (13). They showed that intraerythrocytic 2,3-diphosphoglycerate and P_i could be detected and measured and, furthermore, that intraerythrocytic pH could be deduced from a detailed study of the chemical shifts (see the section on Structure of High-Resolution Spectra). This work was extended in 1974 by Henderson et al., who also detected the ATP in human

Table 8.1. Chemical Shifts of Metabolites Studied by TMR

Metabolite	*Chemical Shift (ppm)*[a]
Adenosine monophosphate	− 6.7
3-P of diphosphoglycerate	6.6
Fructose-6-phosphate	5.6
2-P of diphosphoglycerate	5.5
P_i	5.0
PCr	0.0
γP of ATP	− 2.7
βP of adenosine diphosphate	− 3.1
αP of adenosine diphosphate	− 7.3
αP of ATP	− 7.8
Nicotinamide-adenine dinucleotide	− 8.3
βP of ATP (Mg bound)	−16.5
βP of ATP	−18.2

[a]In live tissue at about pH 7 referenced to PCr.

erythrocytes (23). In the same year, Hoult et al. (14) showed that signals from ATP, PCr, and P_i could be obtained from muscle samples. These results led to the development of techniques for the study of in vivo chemistry.

Table 8.1 gives a list of the chemical shifts of the commonly detected phosphate metabolites. The assignment of the chemical shifts to specific phosphorus atoms is achieved by various methods, including model studies on pure substances and fitting peak heights to the metabolite concentrations known from other methods.

A detailed survey of the many examples of in vitro studies or of various aspects of metabolism by NMR is outside the scope of this chapter. Readers are referred to Table 8.2 for a series of examples and to the various reviews (e.g., References 5-11) on the subject. A few studies will now be highlighted because they provide a basis from which to discuss further applications.

Phosphate Metabolism

The major applications of NMR to in vivo chemistry concern phosphate metabolism. This section gives a simplified account of the processes involved (see Reference 80 for a more comprehensive account).

ATP is the key metabolite (the currency) in cell energetics. Its role is shown schematically in Figure 8.14. When energy is required for chemical, osmotic, or mechanical work, ATP is converted into adenosine diphosphate (ADP) and P_i:

$$ATP^{4-} + H_2O \rightleftharpoons ATP^{3-} + P_i^{2-} + H^+. \quad \Delta G^0 = -7000 \text{ cal mole}^{-1} \qquad [5]$$

The released energy comes from the decrease in the electrostatic interactions, which occurs when the terminal phosphate bond is cleared, and is used to "drive" other chemical reactions, which, in turn, achieve the desired effect. ATP is then regenerated by glycolysis and oxidative phosphorylation in the cytoplasm and mitochondria.

The methods of ATP regeneration, anaerobic glycolysis, and oxidative phosphorylation are summarized in Figure 8.15. The first stage is the conversion of glucose, which may have been derived from glycogen or an other polysaccharide, to pyruvate in the cytoplasm. The second stage is the oxidation of pyruvate to either lactate or carbon dioxide and water. The former process is rapid and can occur under anaerobic conditions. However, it produces lactate as an end-product, which has to metabolize elsewhere (in the liver) and makes the cell acidic. The second process, which proceeds via the Krebs cycle in the mitochondria, is more efficient. The ATP so produced is then transported to wherever it is required.

A further high-energy phosphate detectable by NMR is PCr. ADP, ATP, and PCr are in a rapid equilibrium catalized by creatine kinase:

$$ATP^{4-} + Cr + H^+ \rightleftharpoons ADP^{3} + PCr. \qquad [6]$$

Consequently, PCr acts as a "fast-access" reservoir for the reconversion of ADP into ATP. This reaction, unlike glycolysis and oxidative phosphorylation, stores rather than produces energy.

It has long been known that ATP and ADP exist in the cytoplasm as magnesium complexes. This knowledge has been confirmed by the NMR spectra from living tissue (14). As can be seen from Table 8.1, the chemical shifts of the phosphorus nuclei change when they bind to magnesium; as expected, there is a low field shift, especially for the β-phosphorus. The chemical shift of the β-ATP peak in the spectra from animals corresponds to at least 95% of the detected ATP being bound to magnesium (15,54,64).

In Vitro Examples

Since about 1977, work on perfused organs has been successfully carried out by the Oxford group and several other groups in the United States. Results from this work give an indication of the enormous potential of

Table 8.2. Examples of in Vitro and in Vivo Biochemical Studies by NMR

Organ	*Nature of Experiment*	*Nucleus*	*References*
Brain			
Guinea pig	Perfused studies	^{31}P	24
Rat	Comparison of in vivo and freeze-clamp data	^{31}P	25,26
Rat	Perfused studies	^{31}P	27
Rat	Surface-coil demonstration	^{31}P	19
Heart			
Mouse	Ischemia and recovery	^{31}P	28
Rabbit	Detection of regional ischemia	^{31}P	29
Rabbit	General physiology	^{31}P	30
Rat	Initial studies of ischemic cardiac tissues	^{31}P	31
Rat	Studies of working perfused hearts	^{31}P	32-34
Rat	Creatine kinase kinetics	^{31}P	35
Rat	Acidosis in ischemic heart	^{31}P	36,37
Rat	PCr, and so on, concentration as function of cardiac cycle	^{31}P	38
Rat	Study of pharmacologic treatment of myocardial infarction	^{31}P	39
Kidney			
Rabbit	Studies of ischemia	^{31}P	37
Rat	Model transplant studies	^{31}P	40
Liver			
Mouse	Rapid ATP assay	^{31}P	41
Mouse	Effects of ethanol on alanine metabolism	^{13}C	42
Rat	Rate of lactate production	^{31}P	43
Rat	Cyanide intoxication	^{31}P	44
Rat	Fructose metabolism	^{31}P	45,46
Rat	TMR demonstration and effective ischemia	^{31}P	18

Table 8.2 (*Continued*)

Organ	*Nature of Experiment*	*Nucleus*	*References*
Muscle			
Chicken	Dystrophic studies	^{31}P	47,48
Frog	Initial measurements	^{31}P	14
Frog	Structural changes in myosin	^{31}P	49
Frog	Metabolic studies	^{31}P	50
Frog	Fatigue and energetics	^{31}P	15,51
Frog	pH measurements	^{31}P	51,52
Frog	Activity of creatine kinase	^{31}P	35,53
Frog	Mg^{2+} concentration	^{31}P	54
Mouse	Basic ^{13}C study	^{13}C	55
Mammalian	Normal and Duchenne dystrophy	^{31}P	56-60
Mammalian	Compartmentation	^{31}P	61,62
Mammalian	Unexpected compartmentation	^{31}P	62-64
Cells			
E. coli	ATPase kinetics in aerobic conditions	^{31}P	65
E. coli	pH measurements	^{31}P	66
E. coli	Glucose metabolism	$^{13}C, ^{31}P$	67-69
Erythrocyte (human)	Spin-echo studies	^{1}H	70,71
Liver (rat)	Mitochondrial metabolism	^{31}P	72,73
	Glyconeogenesis in euthyroid and hyperthyroid rats	^{13}C	74,75
Tumor cells (various)	T_1 studies	$^{1}H, ^{31}P$	76-79

NMR in the study of metabolism. The following examples are chosen to illustrate the range of applications.

Cardiac Physiology

The first organs to be studied were perfused rat hearts (see References 31 and 32). The hearts were placed in a cylindric sample tube inside a conventional high-resolution NMR spectrometer and were perfused using Langendorff or other standard methods (see Reference 5 for a review of

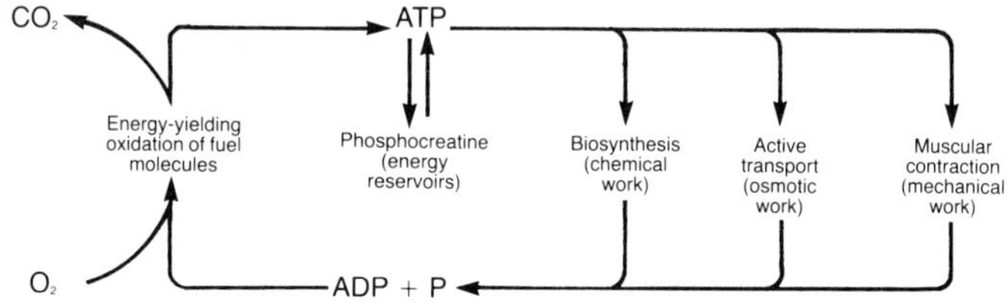

Figure 8.14. ATP is a key metabolite in bioenergetics. The above scheme shows its relationship to ADP, PCr, and P_i.

the methods used) and electrically paced. As expected, the spectra show peaks from the phosphate metabolites present in the heart muscle and indicate a pH of 7.2. Various effects have been studied by these methods.

The metabolic consequences of global ischemia have been extensively studied (27-30). When the beating heart is deprived of oxygen, spectra stored at various time intervals show that the PCr level drops rapidly to maintain the ATP level, and the P_i level increases. The chemical shift of the latter changes as the pH in the cell decreases due to the buildup of lactate; the pH changes from 7.2 to about 6.4 during ischemia (32). On resumption of oxygen supply, the recovery of the heart can be monitored. The extent of the recovery of the heart depends on the duration of the

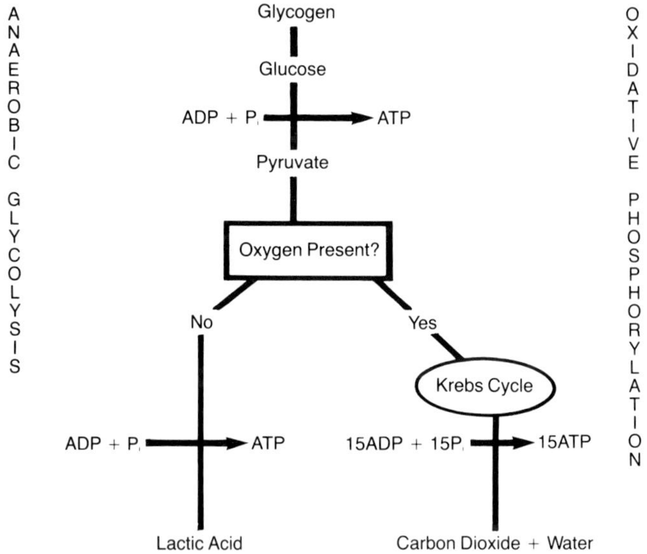

Figure 8.15. Two pathways for the regeneration of ATP from either stored reserves (glycogen) or food (glucose).

ischemic period and the temperature. Hollis et al. (29,30) have studied the effect of arresting the heart with potassium chloride solution before the period of ischemia. Administering potassium chloride solution reduces the decline in pH and ATP concentration during the ischemic period and enhances the ATP recovery when oxygen is restored. Furthermore, if the timing of the data collection is carefully synchronized with the heartbeat, it is possible to follow the relative concentrations of PCr, ATP, and P_i during the course of the cardiac cycle (38).

Myocardial Pharmacology

An interesting recent study has examined the effect of the drugs verapamil and chlorpromazine on the myocardium during periods of ischemia (39). From the ^{31}P spectra obtained on perfused rabbit hearts, Nunnally and Bottomley were able to show that pretreatment of hearts with both drugs preserved phosphate-containing metabolite levels in the ischemic zone after ligation. By inducing coronary vasodilation, verapamil provides virtually complete maintenance of metabolite levels; chlorpromazine, which has no apparent vascular effect, is effective in preventing phospholipid depletion, although not as remarkably well as verapamil. Management of ischemia with verapamil after ligation produced a 1.7- to threefold improvement in PCr levels. This effect represents a return to 50% to 70% of normal levels of this key reservoir of cellular energy. These results demonstrate the value of surface-coil NMR for the quantitative evaluation of the efficacy of drugs used in the management of heart disease.

Transplant Viability

^{31}P NMR studies have been carried out on the kidney. In one study of particular importance for the chemical use of NMR, Sehr et al. (37,40) studied the effect of ischemic-reflow cycles on rat kidneys. They concluded that the reliability of kidneys for transplantation might be assessed by measuring the intracellular pH (via the chemical shift of P_i); the ability to maintain a high pH would indicate a kidney with minimal ischemic damage. In later work, it has been shown that the buffering capacity of the flushing solution produced significant improvements in animal survival studies.

Reaction Kinetics

The previous results have been concerned with the equilibrium concentrations of metabolites more than with the reaction kinetics. By using the saturation-transfer technique, it is possible to apply NMR to studying reaction rates (81). In this technique, the resonant absorption from one site is saturated (in effect, magnetically labeled), and, provided that the reaction has a rate constant greater than the reciprocal of the T_1 of the exchanging phosphate group, saturation is transferred to the other site, resulting in a decrease in signal intensity.

This technique has been applied to the phosphocreatine kinase reaction in frog anaerobic gastrocnemius muscle (53). The resonance due to the γ-phosphate of ATP was saturated and the change in intensity of the PCr line studied. From the results, it can be shown that both the forward (PCr $\rightarrow$ ATP) and reverse fluxes are about 1.6 $mMsec^{-1}$, confirming that the reaction is close to equilibrium. When the muscle is made to work (by electrical stimulation), the forward flux stays at 1.6 $mMsec^{-1}$, whereas the reverse drops to 0.85 $mMsec^{-1}$, giving a net breakdown of PCr of 0.75 $mMsec^{-1}$, which shows that although the reaction is now not at equilibrium, the level of ATP will be maintained. The regulation in this section comes from the ADP produced from the subsequent usage of ATP: A 1% drop in ATP (from 4 to 3.96 mM) produces a threefold increase in ADP (from 20 to 60 μM), which, in turn, drives the phosphocreatine kinase reaction forward (51).

Other muscles, such as myocardium, have also been studied, but here the results are more complex due to competing reactions (74).

Metabolic Pathways

All the previous examples have used ^{31}P NMR. Elegant results have also been obtained with ^{13}C NMR. Due to major experimental problems, for example, low sensitivity, low natural abundance, and the need for heteronuclear decoupling, no in vivo work has been attempted using ^{13}C, until recently. The power of the technique is so strong that it must only be a matter of time before in vitro experiments appear (Fig. 8.6). The following example from the work of Cohen et al. (74) is included to illustrate the potential of the methodology.

Hepatocytes were studied using ^{13}C-labeled glucose, glycerol, and alanine. With glycerol as the substrate, labeling of α-glycerophosphate and glucose could be observed. In addition, the distribution of the ^{13}C label among the various carbon positions in the glucose molecule could be detected. Initially, the ^{13}C appeared at the C-1, C-3, C-4, and C-5 positions. The progress of the label enabled detailed metabolic pathways to be deduced.

The simplicity of the ^{13}C technique should be contrasted with the complexity of the separation techniques necessary to study such processes using ^{14}C as the label. For ^{13}C NMR work, no extraction or degradation, for instance, is necessary. Provided that sufficient concentration of ^{13}C can be achieved (i.e., the molecule studied must be enriched to have a ^{13}C concentration in the millimolar range), the intact system can be studied.

The experimenter has thus the flexibility of following a reaction in a single animal through many stages. The cost of enrichment for animal studies will be high, but the simplicity of the experiment and its nondestructive nature may make the total experimental cost comparable with that of existing techniques. Many further examples of ^{13}C NMR are to be found in the work of Shulman and Cohen (82).

Muscular Dystrophy

Examples have appeared in the literature of ^{31}P spectra taken from human biopsy samples. Burt et al. reported significant differences between the spectra of normal and diseased human quadriceps muscles (58-60). Subsequent work showed that muscles in Duchenne dystrophy do not contain glycerol-3-phosphorylcholine and that their total phosphorus content is lower. Occasionally, Duchenne muscles, especially those with no evidence of X-linked origin, contain normal phosphate levels but no glycerol-3-phosphate. These results are encouraging both from the standpoint of NMR's ability to study diseased as well as healthy tissues and from the standpoint of patients' safety, because the measurements, unlike biopsies, are harmless.

Comparison of ^{31}P NMR Metabolite Measurements with Those of Other Methods

Conventional methods for measuring the concentration of phosphate metabolites in living tissue consist of obtaining a specimen, for example, by growing it in vitro or taking a biopsy, and rapidly freezing it to stop all chemical reactions. The sample is subsequently warmed up and subjected to chemical analysis. How closely the concentration obtained by such methods relates to the concentrations in the original sample is often questionable. NMR, on the other hand, makes its measurements without perturbing the system. It is therefore relevant to compare the values obtained by these two methods.

The major difference between the two methods is that freeze-clamping measures the total metabolite concentrations, whereas NMR measures only the mobile components. The techniques therefore differ with respect to the way they reflect metabolites strongly bound to macromolecules within the cell. In general terms, the techniques agree for ATP and PCr (NMR usually giving slightly higher values, especially for PCr) but disagree for ADP and P_i. NMR reports the latter compounds at a significantly lower level than do freeze-clamp methods.

The reasons for these discrepancies are as yet uncertain; however, the following tentative explanations can be postulated. The ADP measured by NMR is that which is free in the cytoplasm. In muscle, it is assumed that a substantial amount of ADP is bound to the protein of the myofilaments, which may explain the different values reported for that tissue. A second source of discrepancy is that during the cooling period of a freeze-clamp experiment and the subsequent extraction, high-energy phosphates must suffer unavoidable breakdown, which will yield high levels for ADP and P_i. Thirdly, appreciable amounts of the metabolites are found in the mitochondria, where they are not likely to be detected by NMR. Spectra from concentrated suspensions of rat liver mitochondria at 0°C do, however, show ^{31}P signals from ATP and ADP inside and outside the mitochondria (75). The signals are resolved because the internal ATP/ADP pool is predominantly bound to the metal ions, whereas the external pool is not. As discussed previously, complexing of ATP and ADP to metal causes a marked shift in the β-phosphorus resonance. Two P_i signals are also seen, indicating a difference of pH, that of the mitochondria being about 0.5 units higher. The signals from the mitochondrial metabolites are significantly broader than those originating externally. The consensus is, however, that the phosphate signals seen by NMR come from metabolites in the cytoplasm (51).

Compartmentation is an area of research where ^{31}P NMR will probably make important contributions in the future. Work on frog muscle indicates that the concentration of free ADP is only about 20 μM (51), and other results indicate similar values for skeletal muscle in general. The concentration of P_i measured by NMR in a variety of muscles is only about 1-1.5 mM, and the concentrations in kidney, liver, and brain will not be higher. Typical ATP levels measured by NMR are 4 mM. There are some important ramifications of the observation that free ADP and P_i are low in healthy tissue.

The phosphorylation potential ($[ATP]/[ADP]\,[P_i]$) will be much higher than expected from freeze-clamp data. Also, because the concentration of free adenosine monophosphate (AMP) is linked via the adenylate kinase reaction to that of ATP and ADP, it follows that it also is low, about 0.1 μM (a factor of 100 compared with freeze-clamp data). AMP and ADP have important roles in the regulation of many biochemical processes; consequently, the rates and mechanism of many key reactions in vivo may have to be reevaluated. A more detailed discussion of this topic is given by Gadian (5).

RECENT IN VIVO STUDIES

Spectrometers capable of true in vivo studies have only recently become available, and to date only small systems with bores of up to 200 mm have been used. The next sections outline work that has been carried out at Oxford Research Systems, in collaboration with several groups.

Tissue Characterization

The first in vivo high-resolution proton spectrum of human tissue (forearm) was shown at the 1980 Nuclear Magnetic Resonance Imaging Symposium, Nashville, Tennessee (83); a similar spectrum is shown in Figure 8.5. The spectra, which have line widths of about 20 Hz, were recorded at 80.3 MHz on a TMR 32 spectrometer using a 4-cm diameter surface coil. The spectra obtained from human muscle tissue are significantly variable, both from limb to limb and from individual to individual. However, they all consist basically of two main features that are assignable to the $(CH_2)_n$ resonances of the lipid chains of the depot fat present in the limb and tissue water. With suitable resolution enhancement, it is possible to resolve more lines, which can be tentatively assigned to the various protons in the fatty acid chain, for example, the terminal methyls, the methane protons, and those bonded to unsaturated carbons. More detailed work on the assignment of these various resonances is under way.

The relative areas of the water-to-"fat" resonances are believed to be related to proportions of muscle fibers to depot fat present in the "sample" studied. Such measurements, which are rapid, provide a new method for studying pathologic processes in muscular dystrophy (which will complement the ^{31}P studies mentioned previously), and the ability to estimate the fat content of any tissue is clearly of importance in conditions that involve fatty degeneration.

The knowledge that the proton signals from human muscle consist of many resonances with very different T_1 values, and not, as has been frequently assumed, simply of water with a single T_1, must be borne in mind when interpreting the origins of "contrast" in NMR images.

^{31}P Studies of the Human Forearm

The recent advances in instrumentation outlined previously have made it possible to obtain high-resolution NMR spectra from a human forearm in a few minutes. Results have been shown for ^1H and ^{13}C in Figures 8.5 and 8.6. The arm is immobilized in an accurate geometric relation to the $\mathbf{H}_1$ coil by a plastic cradle and a hand grip, which also serves to measure grip strength. Figure 8.16 is the ^{13}P spectrum of a human forearm showing the expected resonances from ATP, and so on. Because of the proximity of bone and the inhomogeneity of the $\mathbf{H}_0$ and $\mathbf{H}_1$ fields, convolution dif-

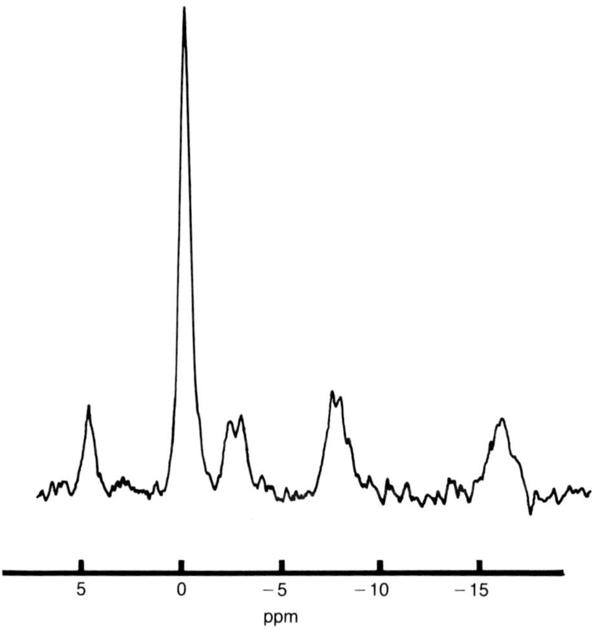

Figure 8.16. ^{31}P TMR spectrum of a human forearm taken at 32.5 MHz in a 20-cm spectrometer. The region examined is approximately 4 cm in diameter and 2 cm deep, situated 6 cm distal to the medial epicondyl; it includes the fleshy parts of the palmaris longus and flexor carpi radialis muscles. Eight hundred scans were taken at 2-sec intervals, and the spectrum was enhanced by convolution differencing.

ferencing is necessary; however, the spectrum is sufficiently well resolved to show phosphorus-phosphorus homonuclear spin coupling. The peak between P_i and PCr is inconspicuous, but because it was present in all seven subjects examined, a measurable quantity of phosphodiester must be present (48). Preliminary calculations on the position of the P_i peak indicate an internal pH of 7.1.

Spectra from the arm can be obtained in a few minutes. It is thus possible to study time courses, for example, of the effect of restricting the blood supply with a tourniquet applied to the upper arm, as is shown in Figure 8.17 (83), and the effects of exercise in the aerobic or anaerobic conditions. Studies of this type are providing results that will be published elsewhere (84).

As was mentioned above, the convolution-difference procedure is used to improve the resolution of the spectra produced when TMR is used; accordingly, the problem of absolute calibration is approached with great

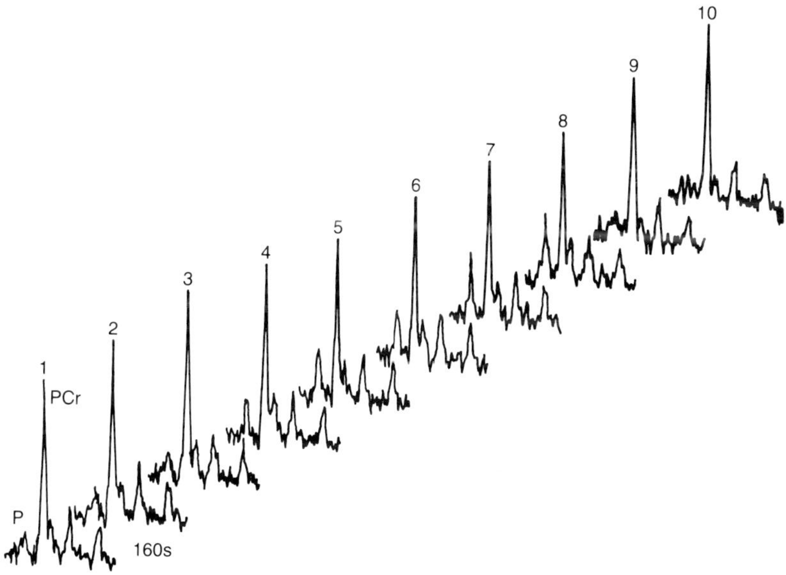

Figure 8.17. Series of ^{31}P spectra of the human arm taken at 160-sec intervals. A tourniquet was applied at the end of spectrum 3 and released at the end of spectrum 8. Note the changes in the relative concentrations of P_i, PCr, and the constant level of ATP.

caution because it can no longer be assumed that the peak areas are proportional to the concentrations of metabolites in millimoles per kilogram of muscle. Ultimately, the absolute calibration rests on a comparison with chemical analyses of samples of human muscle obtained either by needle biopsy or open operations. Fortunately, the strengths and weaknesses of chemical analysis and of TMR complement each other, so that reasonably accurate absolute calibration is possible. Figure 8.18 shows the effect of complete occlusion of the circulation for 58 min followed by a recovery period. The fall in PCr and, especially, the rapid and relatively large rise in P_i show, among other things, that TMR is a powerful tool for detecting anoxia. The relative constancy of $[PCr + P_i]$, combined with the observation that no other peaks altered detectably in area, shows that the saturation factors employed (1.22 for P_i, 1.42 for PCr) were approximately correct, and the mean value, 32.15 mmol/kg, is entirely consistent with the results of chemical analysis.

The most important physiologic result to merge from our studies so far is that even using the relatively fast technique of needle biopsy (approximately 6 sec to freezing), important artifacts are introduced. An appre-

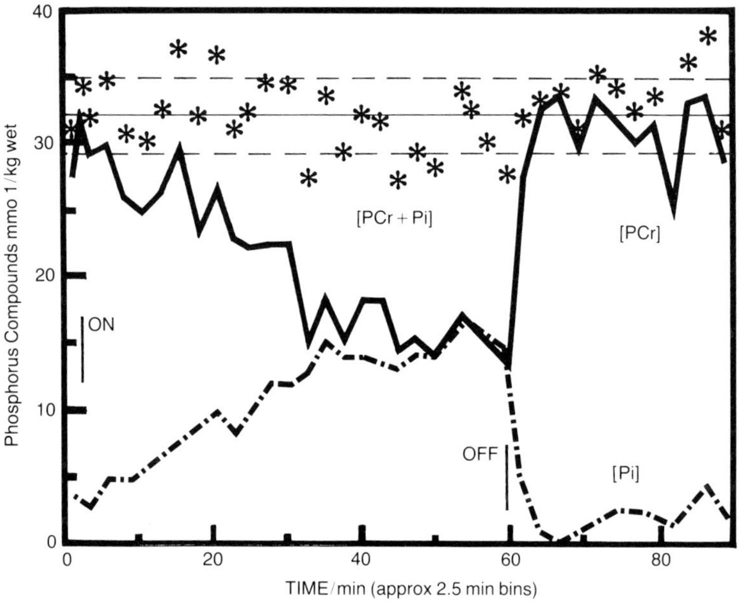

Figure 8.18. Changes in concentration of PCr and P_i during and after 58 min of complete arterial occlusion. The stars indicate the sums of [PCr] and [P_i], with their means ± SD.

ciable quantity of PCr is broken down, and the estimates of P_i and of creatine (Cr) are much too high. This disclosure is important not only in itself but also because the concentration of ADP is proportional to Cr/PCr and the concentration of AMP is proportional to $(Cr/PCr)^2$. Both ADP and AMP have been implicated in theoretical schemes concerning the regulation of metabolism, so it is important to have an accurate estimate of them.

FUTURE APPLICATIONS OF HIGH-RESOLUTION NMR IN MEDICINE

NMR is a harmless, noninvasive technique that provides information not previously attainable by convenient or acceptable means, or simply not attainable. How will these data be used, and what impact will they have? The answers probably lie in the following areas:

Peripheral Vascular Disease

This is the simplest area to start with because the magnet requirements are least demanding, and it will be the first area in which clinical trials will take place. Trials are scheduled to start in Oxford and Philadelphia early in 1981. The major benefits here should be in the quantification of ischemia in the lower limbs and the monitoring of drug effects and efficacy. Much new information will be forthcoming about the various muscular dystrophies, their prognosis, and the possibility of early diagnosis.

Cerebral Infarctions

The brain is notoriously difficult to study. As can be seen from Figure 8.19, spectra from the brain can easily be obtained and its metabolic state

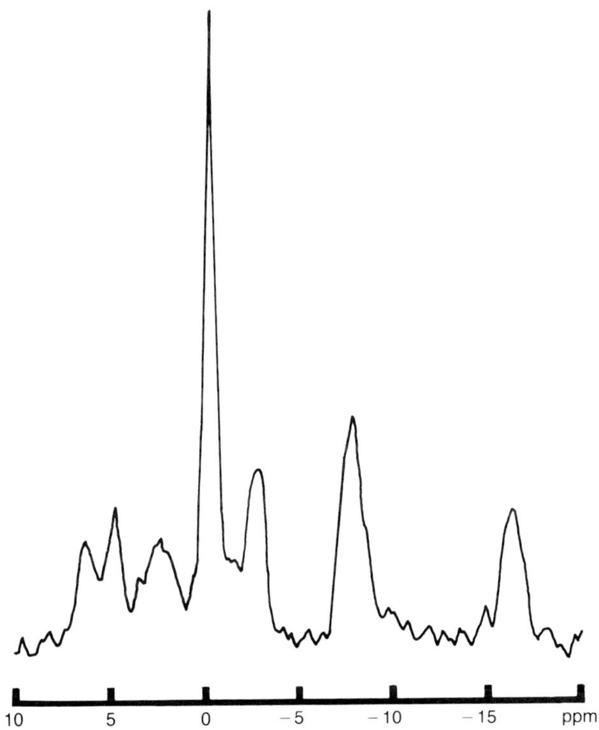

Figure 8.19. ^{31}P spectrum at 32.5 MHz of the brain of an anesthetized rat obtained with a 20-mm surface coil and a 40-mm sensitive volume. The spectrum results from the average of 1024 pulses.

monitored. The availability of such primary data should enable essential, and often difficult, decisions to be made more reliably and sooner than at present. X-ray evidence of brain damage is based essentially on edema, which is a secondary effect and frequently does not occur until after the event.

Transplant Viability

As we discussed in the previous section on this topic, valuable information about organ-preservation techniques has already been obtained. Screening of human kidneys before transplantation has already been started by Radda's group in Oxford (85), with a view to establishing whether there is a correlation between "health" of kidneys as measured, for instance, by pH, and their successful function in recipients. Whether NMR can distinguish between subsequent failure due to ischemia and that due to infection remains to be investigated.

Myocardial Ischemia

Studying infarctions in the heart by TMR techniques will not be easy due to the movement of the organ. These technical problems should not be insurmountable, and their solution should give access, without the risk of present needle biopsy techniques, to valuable information about the health of the myocardium and its response to therapy.

This list could be endless; for example, the work discussed in the section on Recent in Vivo Studies will have implications for sports medicine, the study of liver and kidney disorders and, most importantly, the study of the chemistry of cancer. The detection of tumors via proton spin-lattice relaxation times was the first proposed application of NMR in medicine (76). In vitro ^{31}P studies by Koutcher and Damadian (77) showed high levels of carbohydrate and P_i. Preliminary in vivo results show also that animal tumors have unusually high levels of these components and question the presently held views about the low pH in such cells (86). The future is wide open!

The purpose of this chapter has been to provide the background and a few examples of the role of NMR in the study of in vivo chemistry. Whether NMR imaging will reveal new information not seen on computed tomograms or whether its resolution can be made superior remains to be seen, as does how inherently safe the technique is. What I hope has been shown is that high resolution permits the safe, noninvasive study of in vivo chemistry, that is, that it adds a new dimension to clinical medicine. What impact the unique abilities of NMR to measure in vivo metabolite concentrations and reaction rate, to characterize tissue, and to monitor drug efficacy will have on medicine and disease management in the future also remains to be seen. The impact will certainly be substantial, and the future looks exciting.

ACKNOWLEDGMENTS

I gratefully acknowledge the assistance of G.K. Radda and D.G. Gadian, Department of Biochemistry, Oxford University; J.M. Dawson and D.R. Wilkie, Department of Physiology, University College, London; and my colleagues at Oxford Research Systems Limited in the preparation of this chapter.

REFERENCES

1. Lynden-Bell RM, Harris RK: *Nuclear Magnetic Resonance Spectroscopy.* London, Nelson, 1969.
2. Dwek RA: *Nuclear Magnetic Resonance in Biochemistry.* Oxford, Clarendon, 1973, p 395.
3. Shaw D: *Fourier Transform NMR Spectroscopy.* Amsterdam, Elsevier, 1976.
4. Knowles PF, Marsh D, Rattle HWE (eds): *Magnetic Resonance of Biomolecules.* London, John Wiley & Sons, 1976.
5. Gadian DG: *NMR and Its Applications to Living Systems.* Oxford, Oxford University Press, 1981.
6. Shulman RG (ed): *Biological Applications of Magnetic Resonance.* New York, Academic Press, 1979.
7. Williams RJP, Andrew ER, Radda GK (eds): NMR of intact biological systems. *Philos Trans R Soc London Ser B* 289:379–559, 1980.
8. Brown IR, Gadian DG, Garlick PB, et al: in Dutton L, Leight J, Scarpa A (eds): *Frontiers of Biological Energetics: Electrons to Tissues.* New York, Academic Press, 1978, pp 1341–1351.
9. Radda GK, Seeley PJ: Recent studies on cellular metabolism by nuclear magnetic resonance. *Annu Rev Physiol* 41:749–769, 1979.
10. Burt CT, Cohen SM, Barany M: Analysis of intact tissue with ^{31}P NMR. *Annu Rev Biophys Bioeng* 8:1–25, 1979.
11. Garlick PB, Radda GK: NMR and its applications to metabolic studies. *Tech Metab Res* B216:1–24, 1980.
12. Griffiths JR, Iles RA: Nuclear magnetic resonance: A 'magic eye' on metabolism. *Clin Sci* 59:225–230, 1980.
13. Moon RG, Richards JH: Determination of intracellular pH by ^{31}P magnetic resonance. *J Biol Chem* 248:7276–7278, 1973.
14. Hoult DI, Busby SJW, Gadian DG, et al: Observations of tissue metabolites using ^{31}P nuclear magnetic resonance. *Nature (London)* 252:285–287, 1974.

15. Dawson MJ, Gadian DG, Wilkie DR: Contraction and recovery of living muscle studied by ^{31}P nuclear magnetic resonance. *J Physiol (London)* 267:703–735, 1977.

16. Wehrli FW, Wirthlin T: *Interpretation of Carbon-13 NMR Spectra.* London, Heyden & Sons, 1976.

17. Hoult DI, Richards RE: The signal-to-noise ratio of the NMR experiment. *J Magn Reson* 24:71–85, 1975.

18. Gordon RE, Hanley PE, Shaw D, et al: Localisation of metabolites in animals using ^{31}P TMR. *Nature (London)* 287:736–738, 1980.

19. Ackerman JJH, Grove TH, Wong GG, et al: Mapping of metabolites in whole animals by ^{31}P NMR using surface coils. *Nature (London)* 283:167–170, 1980.

20. Bendel P, Lai CM, Lauterbur PC: ^{31}P spectroscopic zeugmatography of phosphorus metabolites. *J Magn Reson* 38:343–356, 1980.

21. Bottomley PA: Localised NMR spectroscopy by the sensitive point method. 1981, In press.

22. Campbell ID, Dawson CD, Williams RJP, et al: Resolution enhancement of protein PMR spectra using the difference between a broadened and a normal spectrum. *J Magn Reson* 11:172-181, 1973.

23. Henderson TO, Costello AJR, Omachi A: Phosphate metabolism in intact human erythrocytes: Determination by phosphorus-^{31} nuclear magnetic resonance spectroscopy. *Proc Nat Acad Sci USA* 71:2487–2490, 1974.

24. Nishikawa H, Gujii T, Yamada S, et al: ^{31}P NMR study of perfused brain slices of guinea pig. *J Biochem* 87:663–666, 1980.

25. Chance B, Nakase Y, Bond M, et al: Detection of ^{31}P nuclear magnetic resonance signals in brain by in-vivo and freeze-trapped assays. *Proc Nat Acad Sci USA* 75:4925–4929, 1978.

26. Chance B, Radda GK, Seeley PJ, et al: ^{31}P NMR of excised and in-situ brain tissue, in *NMR and Biochemistry.* New York, Marcel Dekker, 1979, pp 269–281.

27. Norwood WI, Norwood CR, Ingwall JS, et al: ^{31}P NMR of perfused brain. *J Thorac Cardiovasc Surg* 78:823–830, 1979.

28. Garlick PB, Seeley PJ, Battersby MK, et al: Phosphorus nuclear magnetic resonance studies on perfused heart, in Esmann V (ed): *Regulatory Mechanisms of Carbohydrate Metabolism.* Oxford and New York, Pergamon Press, 1978, pp 297–302.

29. Hollis DP, Nunnally RL, Jacobus WE, et al: Detection of regional ischaemia in perfused beating hearts by phosphorus nuclear magnetic resonance. *Biochem Biophys Res Commun* 75:1086–1091, 1977.

30. Hollis DP, Nunnally RL, Taylor GJ, et al: ^{31}P NMR studies of heart physiology. *J Magn Reson* 29:319–330, 1978.

31. Gadian DG, Hoult DI, Radda GK, et al: Phosphorus nuclear magnetic resonance studies on normoxic and ischaemic cardiac tissue. *Proc Nat Acad Sci USA* 73:4446–4448, 1976.

32. Garlick PB, Radda GK, Seeley PJ, et al: Phosphorus NMR studies on perfused heart. *Biochem Biophys Res Commun* 74:1256–1262, 1977.

33. Jacobus WE, Taylor GJ, Hollis DP, et al: Phosphorus nuclear magnetic resonance of perfused working rat hearts. *Nature (London)* 265:756–758, 1977.

34. Salhany JM, Pieper GM, Wu S, et al: ^{31}P NMR of perfused heart. *J Mol Cell Cardiol* 11:601–610, 1979.

35. Brown TR, Gadian DG, Garlick PB, et al: Creatine kinase activities in skeletal and cardiac muscle measured by saturation transfer NMR, in Dutton L, Leight J, Scarpa A (eds): *Frontiers of Biological Energetics: Electrons to Tissues*. New York, Academic Press, 1978, vol 2, pp 1341–1349.

36. Garlick PB, Radda GK, Seeley PJ: Studies of acidosis in the ischaemic heart by phosphorus nuclear magnetic resonance. *Biochem J* 184:547–554, 1979.

37. Ackerman JJH, Bore PJ, Gadian DG, et al: Studies in perfused organs by NMR. *Philos Transr Soc (London)* 289:425–437, 1980.

38. Fossel ET, Morgan EM, Ingwall JS: Measurement of changes in high energy phosphates in the cardiac cycle using gated ^{31}P NMR. *Proc Nat Acad Sci USA* 77:3654–3658, 1980.

39. Nunnally RL, Bottomley PA: A ^{31}P NMR assessment of pharmacological treatment of myocardial infarction using surface coils. *Science* 211:177–180, 1980.

40. Sehr PA, Radda GK, Bore PJ, et al: A model kidney transplant studied by phosphorus nuclear magnetic resonance. *Biochem Biophys Res Commun* 77:195–202, 1977.

41. McLaughlin AC, Takeda H, Chance B: Rapid ATP assays in perfused mouse liver by ^{31}P NMR. *Proc Nat Acad Sci USA* 76:5445–5449, 1979.

42. Cohen SM, Shulman RG, McLaughlin A: Effects of ethanol on alanine metabolism in perfused mouse liver studied by ^{13}C NMR. *Proc Nat Acad Sci USA* 76:4808–4812, 1979.

43. Iles RA, Baron PG, Cohen RD: The effect of reduction of perfusion rate on lactate and oxygen uptake glucose output and energy supply in the isolated perfused liver of starved rats. *Biochem J* 184:635–642, 1979.

44. Salhany JM, Stohs SJ, Reinke LA, et al: ^{31}P nuclear magnetic resonance of metabolic changes associated with cyanide intoxication in perfused rat liver. *Biochem Biophys Res Commun* 86:1077–1083, 1979.

45. Iles RA, Griffiths JR, Gadian DG, et al: A ^{31}P nuclear magnetic resonance study of fructose metabolism in the perfused rat liver. *Clin Sci* 58:2p, 1980.

46. Iles RA, Griffiths JR, Stevens AN, et al: Effects of fructose on the energy metabolism and acid-base status of the perfused starved rat liver—a ^{31}P nuclear magnetic resonance study. *Biochem J* 192:191–202, 1980.

47. Chalovich JM, Burt CT, Cohen SM, et al: Identification of an unknown ^{31}P nuclear magnetic resonance from dystrophic chicken as L-serine ethanolamine phosphodiester. *Arch Biochem Biophys* 182:683–689, 1977.

48. Chalovich JM, Burt CT, Danon MJ, et al: Phosphodiesters in muscular dystrophies. *Ann NY Acad Sci* 317:649–668, 1979.

49. Barany M, Barany K, Burt CT, et al: Structural changes in myosin during contraction and the state of ATP in the intact frog muscle. *J Supramol Struct* 3:125–140, 1975.

50. Yoshizaki K: ^{31}P NMR studies of phosphorus metabolites in frog muscle. *J Biochem* 29:12–18, 1977.

51. Dawson MJ, Gadian DG, Wilkie DR: Muscular fatigue investigated by phosphorus nuclear magnetic resonance. *Nature (London)* 274:861–866, 1978.

52. Yoshizaki K, Nishikawa H, Yamada S, et al: Intracellular pH measurements of frog muscle by means of ^{31}P NMR. *Jap J Phys* 29:211–235, 1979.

53. Brown T, Chance EN, Dawson MJ, et al: The activity of creatine kinase in frog skeletal muscle studied by ^{31}P NMR. *J Physiol (London)* 267:703–735, 1980.

54. Cohen SM, Burt CT: ^{31}P nuclear magnetic relaxation studies of phosphocreatine in

intact muscle: Determination of intracellular free magnesium. *Proc Nat Acad Sci USA* 74:4271–4275, 1977.

55. Fung BM: Carbon-13 and proton magnetic resonance of mouse muscle. *Biophys J* 19:315–319, 1977.
56. Burt CT, Glonek T, Barany M: Analysis of living tissue by phosphorus-31 magnetic resonance. *Science* 195:145–149, 1977.
57. Burt CT, Chalovich JM, Danon MJ, et al: Phosphorus nuclear magnetic resonance of diseased muscle, in Dutton PL, Leigh JS, Scarpa A (eds): *Frontiers of Biological Energetics*. New York, Academic Press, 1978, pp 1371–1378.
58. Burt CT, Danon MJ, Millar EA, et al: Variations of phosphate metabolites in normal and Duchenne human muscle. *Biophys J* 21:184a, 1978.
59. Barany M, Burt CT, Labotka RJ, et al: ^{31}P NMR of diseased muscle, in Rowland LP (ed): *Pathogenesis of Human Muscular Dystrophies*. Amsterdam, Excerpta Medica, 1977, pp 337–340.
60. Barany M, Chalovich MJ, Burt CT, et al: Nuclear magnetic resonance studies of muscle, in Schotland DL (ed): *Diseases of the Motor Unit*. Boston, Houghton Mifflin, 1981.
61. Seeley PJ, Busby SJW, Gadian DG, et al: A new approach to metabolite compartmentation in muscle. *Biochem Soc Trans* 4:62–64, 1976.
62. Busby SJW, Gadian DG, Radda GK, et al: Phosphorus nuclear magnetic resonance studies of compartmentation in muscle. *Biochem J* 170:103–114, 1978.
63. Burt CT, Glonek T, Barany M: Phosphorus 31 nuclear magnetic resonance detection of unexpected phosphodiesters in muscle. *Biochemistry* 15:4850–4853, 1976.
64. Burt CT, Glonek T, Barany M: Analysis of phosphate metabolites, the intracellular pH and the state of adenosine triphosphate in intact muscle by phosphorus nuclear magnetic resonance. *J Biol Chem* 251:1584–1591, 1976.
65. Brown TR, Ugurbil K, Shulman RG: ^{31}P nuclear magnetic resonance measurements of ATPase kinetics in aerobic Escherichia coli cells. *Proc Nat Acad Sci USA* 74:5551–5553, 1977.
66. Navon G, Ogawa S, Shulman RG, et al: High resolution ^{31}P nuclear magnetic resonance studies of metabolism in aerobic Escherichia coli cells. *Proc Nat Acad Sci USA* 74:888–891, 1977.
67. Ogawa S, Shulman RG, Glynn P, et al: On the measurement of pH in Escherichia coli by ^{31}P nuclear magnetic resonance. *Biochim Biophys Acta* 502:42–50, 1978.
68. Ugurbil K, Brown TR, den Hollender JA, et al: High resolution ^{13}C NMR studies of glucose metabolism in E. coli. *Proc Nat Acad Sci USA* 75:3742–3746, 1978.
69. Shulman RG, Brown TR, Ugurbil K, et al: Cellular applications of ^{31}P and ^{13}C nuclear magnetic resonance. *Science* 205:160–166, 1979.
70. Campbell ID, Dobsin DM, Williams RJP, et al: Pulse methods for the simplification of protein NMR spectra. *FEBS Lett* 57:96–99, 1975.
71. Brown FF, Campbell ID, Kuchei PW, et al: Human erythrocyte metabolism studies by ^{1}H spin-echo NMR. *FEBS Lett* 82:12–16, 1977.
72. Ogawa S, Rottenberg H, Brown TR, et al: High resolution ^{31}P nuclear magnetic resonance study of rat liver mitochondria. *Proc Nat Acad Sci USA* 75:1796–1800, 1978.
73. Ogawa S, Brown TR, Rottenberg H, et al: ^{13}P NMR study of mitochondria. *Biophys J* 21:60a, 1978.

74. Cohen SM, Ogawa S, Shulman RG: ^{13}C NMR studies of gluconeogenesis in rat liver cells: Utilization of labeled glycerol by cells from euthyroid and hyperthyroid rats. *Proc Nat Acad Sci USA* 76:1603–1607, 1979.

75. Cohen SM, Ogawa S, Rottenberg H, et al: ^{31}P nuclear magnetic resonance studies of isolated rat liver cells. *Nature (London)* 273:554–556, 1978.

76. Damadian R: Tumor detection by NMR. *Science* 171:1151–1153, 1971.

77. Koutcher JA, Damadian R: Spectra differences in ^{31}P NMR of normal and malignant tissue. *Physiol Chem Phys* 9:181–187, 1977.

78. Damadian R: Field focusing NMR (Fonar) and the formation of chemical images in man and references therein. *Philos Trans R Soc London Ser B* 289:389–500, 1980.

79. Navon G, Ogawa S, Shulman RG, et al: ^{31}P nuclear magnetic resonance studies of Ehrlich ascites tumour cells. *Proc Nat Acad Sci USA* 74:87–91, 1977.

80. Lehninger AL: *Biochemistry*. New York, Worth Publishing, 1975.

81. Brown TR: Saturation transfer in living systems. *Philos Trans R Soc London Ser B* 289:441–444, 1981.

82. Cohen SM, Shulman RG: ^{13}C studies of gluconeogenesis of rat liver suspensions and perfused rat liver. *Philos Trans R Soc London Ser B* 289:407–441, 1980.

83. Shaw D, Gordon RE, Hanley PE, et al: Topical magnetic resonance studies of the human arm. *J Comput Assist Tomogr*, 1981.

84. Dawson MJ, Edwards RE, Wilkie DR, et al: ^{31}P NMR studies of metabolites in the human forearm. *Physiol Soc*, 1981, In press.

85. Bore PB, Chan L, French M, et al: ^{31}P NMR of isolated human kidney. 1981, In press.

86. Stephens AN, Iles RA, Griffiths JR, et al: NMR studies of tumours. 14th FEBS meeting, 1981, To be published.

9

Imaging of Elements Other Than Hydrogen

David M. Kramer, Ph.D.

GENERAL CONSIDERATIONS

There is an increasing amount of interest in NMR zeugmatographic imaging of endogenous elements other than hydrogen. The potential and practical limitations that one faces will be assessed in general terms for a wide variety of nuclei. Several elements will be discussed in detail because they seem to offer the greatest potential for revealing local differences in physiochemical structure or function without overwhelming practical difficulty. Some discussions of these topics have been published previously (1,2) in lesser detail.

The most useful way to discuss questions of NMR sensitivity, image resolution, and other performance criteria is on a relative scale with respect to protons (^1H). In that way, as the technology improves, the relevance of this discussion will not be lost.

The single most important variable is the signal-to-noise ratio, that is, the overall NMR sensitivity of the isotope in question for the concentration at which it is present in tissue. Table 9.1 provides some NMR data on the most abundant elements in living organisms (3). The relative sensitivity is given for constant frequency for reasons that will be discussed below. Also given in Table 9.1 are the NMR frequencies of the same nuclei in a field of 1 Tesla.

The explanation for choosing a constant-frequency comparison of sensitivity is that the optimum operating frequency will be independent of the element being imaged. Basically, the optimum range of operation will be determined by two competing effects: the intrinsic NMR sensitivity, which increases as frequency to the 3/2 power (4), and losses due to decreasing penetration of radio frequency magnetic fields (5-8). The op-

Table 9.1. NMR Data on Selected Isotopes[a]

Isotope	*NMR Frequency (MHz) at 1 Tesla*	*Natural Abundance (%)*	*Sensitivity at Constant Frequency for Equal Number of Nuclei*	*Spin (units of h/2π)*
^{1}H	42.5759	99.9844	1	1/2
^{2}H	6.53566	0.0156	0.409	1
^{11}B	13.660	81.17	1.60	3/2
^{13}C	10.705	1.108	0.251	1/2
^{14}N	3.076	99.635	0.193	1
^{15}N	4.315	0.365	0.101	1/2
^{17}O	5.772	0.037	1.58	5/2
^{19}F	40.055	100	0.941	1/2
^{23}Na	11.262	100	1.32	3/2
^{25}Mg	2.606	10.05	0.714	5/2
^{27}Al	11.094	100	3.04	5/2
^{29}Si	8.458	4.70	0.199	1/2
^{31}P	17.237	100	0.405	1/2
^{33}S	3.266	0.74	0.384	3/2
^{35}Cl	4.172	75.4	0.490	3/2
^{37}Cl	3.472	24.6	0.408	3/2
^{39}K	1.987	93.08	0.233	3/2
^{43}Ca	2.865	0.13	1.41	7/2
^{57}Fe	1.38	2.245	0.032	1/2
^{63}Cu	11.285	69.09	1.33	3/2
^{67}Zn	2.664	4.12	0.730	5/2
^{85}Rb	4.111	72.8	1.13	5/2
^{87}Sr	1.845	7.02	1.43	9/2
^{127}I	8.519	100	2.33	5/2
^{137}Ba	4.732	11.32	0.556	3/2
^{207}Pb	8.899	21.11	0.209	1/2

[a]From Varian Associates NMR Table, ed 5, 1965.

timum frequency is likely to encompass a broad range. It will ultimately depend on how efficiently one can correct the distortion of the signal due to attenuation in tissue. Related problems occur in other modalities, such as ultrasonic imaging and emission tomography, and have been overcome at least to some degree (9,10).

To observe other nuclei at the same frequency as protons, much stronger fields will be required. For example, ^{31}P requires two and one half times the magnetic field strength needed for protons; carbon or sodium, about four times the strength. To achieve these high fields over large volumes will require superconducting magnets. Thus, the cost of whole-body magnets for NMR imaging of other nuclei will be high, in terms of both the initial investment and the additional technical upkeep compared to conventional electromagnets. Another issue that will have to be dealt with is the risk to a patient in the event of a sudden quenching of a superconducting system. Budinger (11) has shown that the biologic effect of a field change of the order of 1 Tesla/sec is potentially hazardous.

Considerations of such matters, although obviously important, seem premature. It may be that protons also will require superconducting magnets, depending on the optimal frequency.

Returning to the discussion of sensitivity, the next question relates to the expected concentrations of various elements other than hydrogen in the body. Table 9.2 gives some approximate data on the average levels of NMR elements in mammalian tissue. The data in the first column are taken from Bowen (3) and are expressed as parts per million in the dry tissue. The numbers in the second column are calculated on the basis of the natural abundance in Table 9.1. For example, although carbon represents nearly half the dry weight of tissue, almost all carbon is ^{12}C and has no magnetic moment. Only 1% is ^{13}C, which can be observed by NMR. In Table 9.1, only the most abundant isotope of a given element is represented, except for hydrogen and nitrogen.

Actually, it is the level of a given isotope per unit volume in the wet tissue that is important. The data can be converted to volume concentrations by allowing that the tissue is 75% water by either weight or by volume. This is shown in the third column. It should be noted immediately that protons (^{1}H) are two orders of magnitude more concentrated than is the next most common nucleus. This is the determining variable that makes ^{1}H the most attractive element for NMR zeugmatographic imaging.

The combination of the concentration, natural abundance, and the intrinsic relative sensitivity gives the overall scale of NMR sensitivity. Table 9.3 compares the approximate sensitivities for the selected elements in the body. Given in the right-hand column are the estimates of the corresponding spatial resolution for a cubic picture element, that is, the inverse of the cube root of sensitivity. This would be the resolution available in a three-dimensional NMR zeugmatogram with the contrast, signal-to-noise ratio, and performance time fixed because this is the volume of tissue that would induce the same signal strength in the receiver coil. For example, if a particular nucleus is intrinsically one tenth as sensitive as ^{1}H and is present in the body at only one hundredth the concentration of ^{1}H, taking the natural abundance into account, the overall sensitivity is 1000 times less than that of ^{1}H. Therefore, a volume of tissue 1000 times greater will be needed to produce a signal of equal

Table 9.2. Elementary Composition of Mammalian Tissue for Selected Elements (> 1 ppm)

Element	*ppm (dry tissue)*[a]	*NMR Isotope*[b]	*Molar Concentration*[c]
^{1}H	66,000	66,000	99
^{2}H		10	1.5×10^{-2}
B	<2	<2	4.5×10^{-5}
C	484,000	5360	0.10
^{14}N	87,000	87,000	1.6
^{15}N		320	5.9×10^{-3}
O	186,000	69	3.1×10^{-2}
F	500	500	6.6×10^{-3}
Na	7300	7300	7.8×10^{-2}
Mg	1000	100	10^{-3}
Al	<3	<3	3×10^{-5}
Si	120	5.6	4.8×10^{-5}
P	43,000	43,000	0.35
S	5400	40	3.0×10^{-4}
Cl	3200	2400	1.7×10^{-2}
K	7500	7000	4.5×10^{-2}
Ca	85,000	110	6.4×10^{-4}
Fe	160	3.6	1.6×10^{-5}
Cu	2.4	1.6	6.3×10^{-6}
Zn	160	6.6	2.5×10^{-5}
Rb	18	13	3.8×10^{-5}
Sr	21	1.5	1.8×10^{-5}
I	0.43	0.43	$[8.5 \times 10^{-7}]$
Ba	2.3	0.26	4.7×10^{-7}
Pb	4	1	10^{-6}

[a]From Bowen (3), Chapter 5.

[b]ppm × natural abundance in Table 9.1.

[c]Assumes that 75% of tissue mass and volume is H_2O.

strength. For a cubic resolution element, the length of an edge will be 10 times greater. To make comparisons for a slice of fixed thickness, one need only take the inverse square root of the sensitivity figure, thus conserving the "signal volume." Bearing in mind that the difference in concentration between various organs probably will be larger for other nuclei than it is for protons, one might expect greater inherent image contrast. If this is the case, the resolution will be somewhat better than indicated in Table 9.3 for the same level of detectability.

The numbers given in Table 9.3 are meant only as a guide as to which elements will have some immediate potential, which ones are possible

Table 9.3. NMR Zeugmatographic Imaging Sensitivity

Element	*Overall Tissue NMR Sensitivity*	*Linear Resolution for Fixed* S/N *and Performance Time (sensitivity*$^{-1/3}$)
^{1}H	1.000	1.000
^{2}H	6.2×10^{-5}	25
B	7.3×10^{-7}	110
C	2.5×10^{-4}	16
^{14}N	3.1×10^{-3}	6.9
^{15}N	6.0×10^{-6}	55
O	4.9×10^{-4}	13
F	6.3×10^{-5}	25
Na	10^{-3}	10
Mg	7×10^{-6}	50
Al	9×10^{-7}	100
Si	9.2×10^{-8}	220
P	1.4×10^{-3}	8.9
S	1.1×10^{-6}	97
Cl	8.4×10^{-5}	23
K	1.1×10^{-4}	21
Ca	9.1×10^{-6}	48
Fe	5.2×10^{-9}	580
Cu	8.5×10^{-8}	230
Zn	1.8×10^{-7}	180
Rb	4.3×10^{-7}	130
Sr	2.6×10^{-7}	160
I	[2.0×10^{-8}]	[370]
Ba	2.6×10^{-9}	730
Pb	3×10^{-9}	700

candidates for the future as the technology improves, and finally which ones probably will not be practical for imaging. There are a number of caveats to bear in mind. The concentrations are, to some extent, misleading, especially for elements with low concentrations. This is due to contamination from dust and from handling with metal implements. Also, the values presented are averages of many different species and tissue types. Also, certain elements are sequestered by different organs, making the actual elemental composition very different in any one organ than the average. Iodine, which is concentrated in the thyroid, is one well-known example.

It is appropriate at this point to make one other remark concerning sensitivity. All the data shown here refer to the naturally occurring elemental concentrations and isotopic abundances. It is possible to artificially alter the concentrations of a dilute isotope, for example, by feeding isotopically enriched nutrients to laboratory plants and animals (12). This approach is expensive, so it has limited application, but it can conceptually heighten image contrast in useful and interesting ways. Experiments on small animals seem especially feasible because their small size will permit the use of much higher frequencies and fields.

According to the data in Table 9.3, there are, in fact, some clear categories. It seems safe to say that the 10 most promising elements other than ^{1}H are ^{14}N, ^{15}N, phosphorus, sodium, oxygen, carbon, potassium, chlorine, fluorine, and the deuterium isotope of hydrogen, but not necessarily in that order. There is a gap in sensitivity after that. It would be foolish to rule out any possible application for all the elements that are less sensitive than those just mentioned, but perhaps just as foolish to pursue them at this early stage of NMR imaging.

Let us spend the remainder of this chapter in discussing each of the 10 most sensitive elements in greater detail. It should become increasingly clear that for each element, different variables have subtle roles. It will perhaps become clear also that the simple analysis embodied in Tables 9.2 and 9.3 can be quite misleading.

The next consideration in evaluating a given nucleus as a candidate for NMR imaging is an examination of the conventional high-resolution spectrum. From there, one learns the complexity of the spectrum, the bandwidth, line widths and, hopefully, the chemical identities of any chemically shifted peaks, so that one can assess the potential for direct chemical analysis.

Although it is not the purpose of this chapter to review the high-resolution NMR of all possible elements in biologic materials, it will be useful to take a brief look at the available data for nuclei with favorable sensitivity.

PROSPECTS FOR MOST SENSITIVE ELEMENTS

Nitrogen

Nitrogen is one of the most common elements in the body and might therefore be expected to be a good subject for NMR imaging. Although present in two isotopes with nuclear spin, it is almost all ^{14}N. This is most unfortunate because ^{14}N has spin 1 and therefore has broad resonance lines in the best cases and invisible (extremely broadened) in most. Thus, even though the data in Tables 9.1, 9.2, and 9.3 indicate that nitrogen is perhaps the most sensitive element other than hydrogen, in fact, nitrogen may be impractical for whole-body imaging.

There appear to be few published reports on nitrogen NMR for living biologic systems. One experiment was published (12) in which fungus was grown in a ^{15}N-enriched medium. Four distinct resonances were observed after 17 days. These peaks were in the frequency range typical of peptidyl amides. A review (13) of nitrogen NMR of proteins and amino acids is also available. It would seem that the only naturally occurring nitrogen that could be practical for imaging is ^{14}N from free amino acids, ureas, and other low-molecular-weight components because the ^{14}N resonances of peptides are too broad to be seen. The concentrations of these peptides probably represent just a tiny fraction of the nitrogen in the body. Add to this the fact that these signals will be dispersed over 50 to 100 parts per million frequency range, and the prospects look slim. Nitrogen disappointingly turns out to be a perfect example of how one can be misled by ignoring the chemical state of an element and estimating the NMR sensitivity on the basis of average elemental composition alone.

Phosphorus

Among the most exciting prospects of NMR study of any kind on living systems is phosphorus. It appears as a 100% abundant, spin-1/2 isotope and gives relatively uncomplicated, well-resolved, high-resolution spectra with quite specific physiochemical information. Phosphorus NMR has attracted a great deal of interest because of the unique opportunity it offers to monitor nondestructively the levels of adenosine diphosphate and triphosphate (ADP, ATP), phosphocreatine (PCr), inorganic phosphate (P_i), and sugar phosphates in vivo and thereby study metabolic responses almost in real time. The only drawback, again, is sensitivity.

The literature on ^{31}P NMR in living systems is a rapidly growing one, and it will be reviewed here only briefly. Studies on isolated muscle (14,15), heart (16), liver (17), and brain (18) have demonstrated the ability of ^{31}P NMR to show how metabolic processes are affected by various insults, such as hypoxia and ischemia. The concentrations of phosphorus metabolites as measured by NMR are in good agreement with those from more conventional analysis. The obvious advantages are that the system can remain intact and that repeated studies can be performed on the same organ or animal.

Phosphorus NMR has also been used on whole animals through specialized techniques. A mouse was surgically opened, and the tuned coil of the NMR probe was placed around the kidney (19). Rat hearts have been studied in vivo by similar techniques (20). Another experiment, which does not require surgery, used a flat coil in the NMR receiver (21). When such a flat coil is placed up against the surface of the body, it detects signals up to at least one radius depth. Chance and coworkers (22) have studied the brains of living mice. By using a small species, they were able to insert the whole animal into an 18-mm bore NMR tube and position the head at the center of the receiver coil.

NMR zeugmatographic images of the concentration of phosphorus metabolites have been published (23). The sample consisted of two standard sample tubes: a 15-mm outside diameter tube containing an aqueous solution of 0.15 M ATP and 0.7 M PCr at pH 7.12 and a 5-mm outside diameter inner tube containing aqueous 1.6 M P_i at pH 5.72. The sample was imaged at 145.7 MHz in a 0.37 G/cm gradient that was electronically rotated by 6-degree steps in the plane perpendicular to the tube axes. The chemically shifted signals partially overlapped in the gradient; however, they were separable. They can be separated directly (24) or, indirectly, by reversing the magnetic field gradient and applying suitable mathematical analysis of the data. Each projection required averaging of 1024 transient responses to 45-degree radiofrequency pulses applied every 141 msec.

The images, reconstructed from 30 projections by the filtered-back projection (25) method, are two-dimensional maps of the concentration of the chemically resolved phosphorus metabolites one encounters in tissue. In Figure 9.1 are the images of PCr, P_i, and ATP, respectively. The same images are shown in Figure 9.2 displayed on a normalized intensity scale. In this way, the images show the relative concentrations of the metabolites.

Clearly, the signal-to-noise ratio will be worse in intact biologic specimens because the concentrations are more than an order of magnitude lower. Thus, at the present state of technology, whole-body imaging of ^{31}P is possible only with very low resolution. Using the concentrations of phosphorus that have been reported (14) from NMR measurements, the total concentration of NMR-observable ^{31}P is about 50 mM, a factor of 7 less than the average total concentration. The overall NMR sensitivity is therefore actually 2×10^{-4} that of protons. This sensitivity corresponds to a linear resolution length 17 times that of protons.

On the other hand, low resolution might be enough to make certain experiments practical, especially if very specific chemical data are revealed. For example, an ischemic area of myocardium will contain levels of PCr that are greatly reduced and levels of P_i that are greatly elevated compared to those of normally perfused tissue (16). A selective excitation of, for example, only the PCr followed by observation in linear gradients

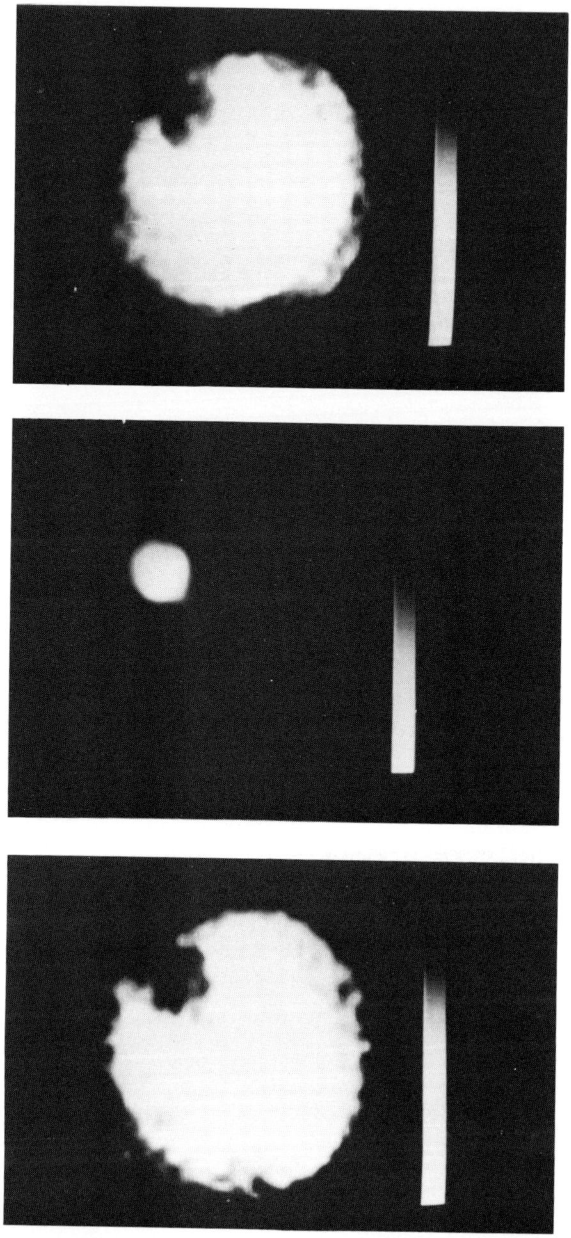

Figure 9.1. Two-dimensional projected NMR zeugmatogram of PCr (top), P_i (middle), and ATP (bottom) reconstructed from 30 projections on a 65 × 65 array. The image intensities in each case are scaled for maximal dynamic range in the display. From Bendel et al. (23) by permission of *Journal of Magnetic Resonance*.

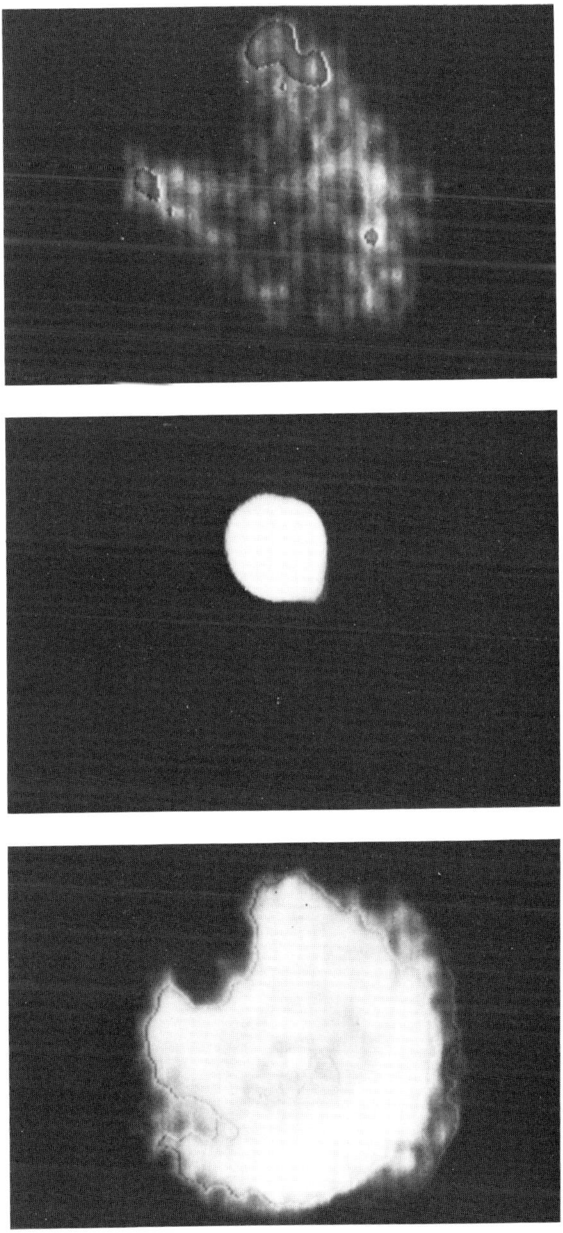

Figure 9.2. Same as in Figure 9.1, except that all three images are on a common intensity-display scale. In this figure, brightness indicates actual relative concentrations. From Bendel et al. (23) by permission of *Journal of Magnetic Resonance*.

could provide a moderate-resolution, low signal-to-noise image capable of quantifying the extent of the ischemic injury. The change in PCr levels would probably be detectable with a $S:N$ ratio of 2 and, therefore, resolution only about 10 times worse than that of protons.

Other trials of ^{31}P NMR imaging have been made. Hoult (26) published an image of a phantom filled with phosphoric acid by a technique he named "rotating frame zeugmatography," which employs linear gradients in the radio frequency field. Two-dimensional Fourier NMR experiments analogous to the Fourier NMR zeugmatography (27) have been done with one magnetic field gradient producing one dimension of spatial information and one of high-resolution NMR spectral information on test samples containing phosphorus metabolites (28).

Also reported are related nonimaging ^{31}P NMR experiments: high-resolution spectra of localized points within larger objects. One experiment (29) used a modification of the sensitive-point imaging method (30) to observe a small region of a perfused, beating heart. Another approach, involving observation of signals from a saddle point in an inhomogeneous static field (31,32), was used to study living rats (33) and a human forearm (34). A surface radio frequency coil, with its greater efficiency than a body-sized coil, could also be used for selectively detecting signals from a small volume. For example, experiments aimed at assessing the ability of various drugs to reduce permanent damage to the heart after ischemia might be designed for study on the anterior left ventricular wall by monitoring the ratio of PCr to P_i or the splitting of the P_i signal due to pH changes (16).

Sodium

As was seen in Table 9.1, sodium is present as one isotope, requires almost four times as much magnetic field, and is 32% more sensitive than hydrogen. It is present in soft tissue in varying degrees, but the average concentration would appear to be about 70 mM. Based on this figure, the overall NMR sensitivity is three orders of magnitude worse (leading to linear resolution about 10 times worse) than that of hydrogen.

In spite of the various technical difficulties, an NMR zeugmatogram based on ^{23}Na has recently been obtained (35). A perfused, beating rat heart, contained in a 20-mm outside diameter tube, was imaged at 95 MHz (82 kG) in a bath of perfusion medium that contained 145 mM sodium. The imaging method employed was a two-dimensional projection-reconstruction sequence combined with planar selection by an orthogonal modulated gradient (36). The data acquisition was synchronized with the heart beat, and two images were formed: one at systole (Fig. 9.3) and

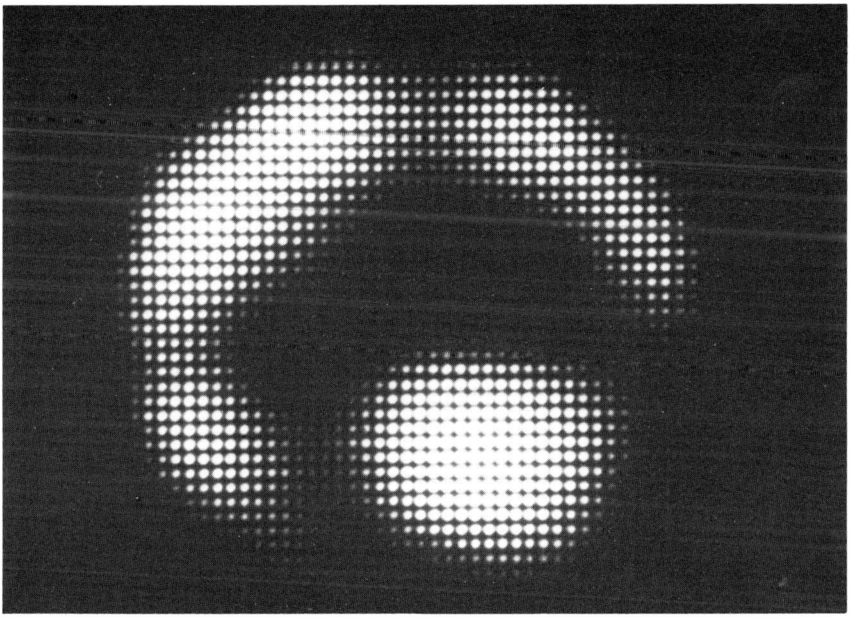

Figure 9.3. Thin-section ^{23}Na NMR zeugmatogram of a perfused, beating rat heart. The image was formed at systole by synchronization. From Delayre and Fossel (35) by permission of *Science*.

another at diastole (Fig. 9.4). The appearance of the images was as a negative because the sodium concentration was higher in the perfusing fluid. Thus, a bright area within the ventricle and another outside the heart reveal the inner and outer heart walls. The images were obtained in 15 to 20 min, although the instrument settings were not optimized.

The conventional high-resolution NMR spectrum of sodium in tissue contains one peak, about 50 Hz wide, comparable to that of dilute, aqueous sodium chloride solutions. Unlike the resonance in simple solutions, the resonance of sodium in tissue is characterized by a single spin-lattice relaxation time T_1 and two spin-spin relaxation times T_2. Possessing a quadrupolar nucleus (spin $> 1/2$), sodium is subject to strong relaxation effects and splittings into side bands. The latter effect has so far not been observed in tissue. The former has been used to advantage to study solution structure and sodium binding in drugs (37a) and enzymes (37b). A review of quadrupolar interaction is beyond the scope of this chapter. A summary of the theory and the relaxation effects in ^{23}Na NMR in tissue can be found in Reference 38. Because the relaxation effects are very sensitive to the state of sodium in solution, sodium T_1 and T_2 may be

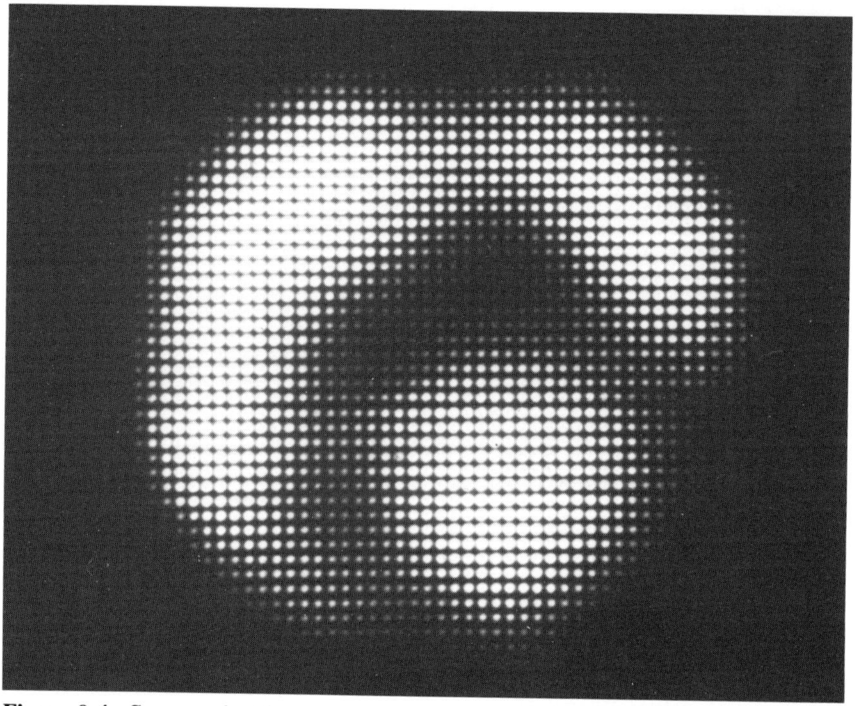

Figure 9.4. Same as in Figure 9.3, except at diastole. From Delayre and Fossel (35) by permission of *Science*.

useful image-contrast parameters for probing function as well as structure. Quadrupolar relaxation effects are also responsible for the fact that only about half the expected signal strength is observed in tissue (39).

With regard to whole-body imaging, there is undoubtedly some potential for ^{23}Na NMR. There are considerable differences in sodium content of organs (40). Brain has more than twice as much sodium as muscle, cardiac and skeletal muscles differ by 10%, liver is 20% to 25% higher than muscle, and kidney has only 20% as much as muscle. Thus, the contrast in ^{23}Na NMR zeugmatography will probably allow imaging at signal-to-noise ratios lower than those of protons and thereby decrease the factor of 10 difference in linear resolution a small amount. It might at least compensate for the fact that about half the expected signal is "invisible."

Oxygen (potassium, chlorine)

Much of what was said of ^{23}Na can be said about ^{17}O, ^{39}K, or ^{35}Cl, except that the signals will be weaker. Potassium and chlorine are present in tissues to somewhat lesser degrees and have less intrinsic sensitivity than does ^{23}Na. Oxygen signals are weak due to low natural abundance. These nuclei are all quadrupolar and can therefore reveal information that protons do not, through quadrupolar relaxation effects, and they have been used as probes to study ion binding in proteins (41) and solution structure (42).

The literature on ^{17}O NMR, for example, is limited, due mostly to practical constraints of low natural abundance. Applications to NMR imaging will be worth pursuing only when a specific advantage is found because probably all the signal will be from water, which is obviously best imaged with ^{1}H.

Carbon

^{13}C is a nucleus that is becoming more and more important for biochemical NMR research. The advantages of this nucleus over protons are larger chemical shifts, no homonuclear coupling (due to low natural abundance), and the same narrow line widths (spin = 1/2). However, imaging of ^{13}C in living systems is made difficult by these large chemical shifts. The high-resolution spectra of proteins have most of the intensity in the aliphatic region, covering a shift range of 50 to 100 ppm (43). Thus, the apparent line width of the ^{13}C-NMR signal will be many times that of protons, thus further reducing the signal-to-noise ratio.

Chemical analysis, except in the crudest fashion, is not a possibility with ^{13}C as it is with phosphorus. Carbon is an element of which there are so many different chemically shifted signals that the spectrum of even a single polypeptide chain of modest molecular weight requires application of numerous special techniques for interpretation.

Even though whole-body, natural-abundance ^{13}C-NMR zeugmatographic imaging is discouraging, there may still be some practical experiments that can be done on small animals at very high fields. Another possibility is combining such an experiment with feeding the animal ^{13}C-enriched nutrients. The main drawback is the high cost of isotopic enrichment.

Fluorine

Fluorine is another example of how one can be misled by basing estimates of NMR imaging performance on average elemental composition without regard for the chemical form. Fluorine is found almost exclusively in teeth and bone. The fluorine signals from these solid materials will be extremely broad.

However, fluorine would be an attractive element to be able to use. It has an intrinsic sensitivity that is almost equivalent to that of protons, its

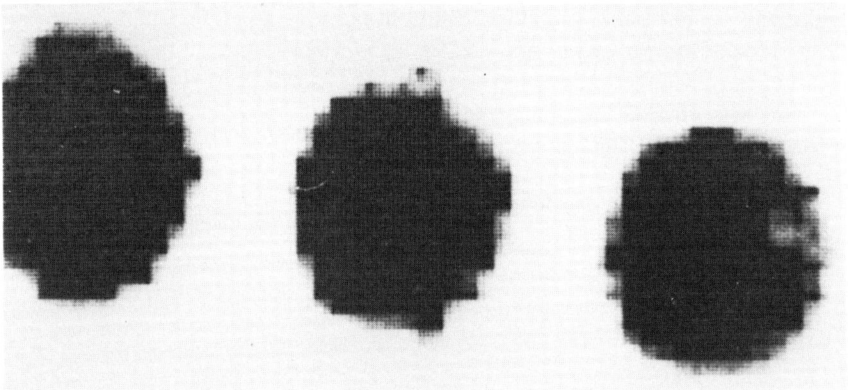

Figure 9.5. Thin-section ^{19}F NMR zeugmatogram of three 1.5-cm diameter cylindric tubes filled with a fluorine-containing compound sometimes used as a blood substitute. From Holland et al. (44) by permission of *Journal of Magnetic Resonance*.

spin is 1/2, so it gives narrow lines, and its magnetogyric ratio is so close to that of protons that the same equipment can be used. In spite of these advantages, it has been overlooked, in large part because of its virtual absence in tissue.

Holland et al. (44) pointed out that the absence of any background ^{19}F signals was an advantage if one could come up with a suitable fluorine-containing compound to introduce as a "tracer." They formed two images of test samples by the multiple sensitive-point method (45). One sample consisted of three glass tubes, 1.5 cm in diameter, containing 1.4 M NaF (see Fig. 9.5). The second sample was a star-shaped pattern of glass tubes, 2 and 6 mm in diameter, filled with perfluorotributylamine (see Fig. 9.6). Each image took 5 min to make and consisted of 64^2 and 128^2 independent points, respectively.

A few potential applications were also pointed out. One particularly interesting fact is that certain fluorocarbons, perfluorotributylamine being one, can be used as artificial blood substitutes in large quantities (44). In this way, images of the cardiovascular system could be made without interfering background.

Deuterium

Deuterium is an isotope that, because of its low natural abundance, will not likely be useful for medical NMR imaging unless a specific advantage over protons is uncovered. Quadrupolar relaxation data is one source for

Figure 9.6. Two-dimensional ^{19}F NMR zeugmatogram of a tube array filled with a fluorine-containing compound sometimes used as a blood substitute. From Holland et al (44) by permission of *Journal of Magnetic Resonance.*

unique data. But other nuclei, with greater sensitivity, are available. Deuterium might have one other interesting difference from protons that is indeed unique. That is, rather strong isotopic effects may cause visible differences in the appearance of NMR images. Hydrogen, for which isotopic differences are relatively large, is the one element that has this potential.

All other things being equal, the ratio of chemical reactivity of two different isotopes of the same element is inversely proportional to the square root of the ratio of their masses. This relationship leads to isotopic abundances in animals that can differ from that in the environment (3). It is to be expected that it may also lead to varying isotopic enrichments between organs. On the other hand, the cell water is being exchanged rapidly with the water in the environment, and the dissociation of hydrogen in water is facile. Thus, the effect will probably appear only in the "non-water" hydrogenated components of the cell. Since these components are only a small fraction of the signal, observation of these differences in natural abundance may prove too difficult, but studies on small animals subjected to deuterated nutrients should not be ruled out. One demonstration of ^{2}H imaging of a test object has recently been published (46).

INDIRECT DETECTION

No discussion of NMR of nuclei other than protons can be said to be complete without mention of indirect detection. Indirect detection refers to observation of an insensitive nucleus through its effects on the spectrum of a sensitive one. In our case, the sensitive nucleus will always be protons, of course. A gain in overall sensitivity can be achieved either because the intrinsic sensitivity of the nucleus of interest (call it x) is lower than that of protons or because each x atom interacts with more than one ^{1}H, increasing the effective concentration, or both. A complete review article on the theoretic and experimental aspects of double-resonance NMR is available (47). The greatest difficulty in these techniques is the large dynamic range difference between the proton signal itself and the relatively small perturbations of interest.

A possibility exists to study the distribution of paramagnetic elements in the body, at very low concentrations, on the basis of their relaxation effects on water-proton relaxation times (48). Such effects were large in dogs and rats for injections of 0.1 mmol of Mn^{2+}/kg of body weight, much less than a lethal dose. These studies were performed to evaluate the use of paramagnetic ions as contrast agents in medical ^{1}H NMR imaging. The relaxation effects of manganese ion on ^{31}P NMR of perfused beating rabbit hearts and the degree to which the effects can be reversed have also been observed (49).

CONCLUSION

Generally speaking, NMR imaging of elements other than ^{1}H will be demanding technologically and the results, with the possible exceptions of sodium, phosphorus and, perhaps, fluorine, of uncertain usefulness. Images of potassium, chlorine, oxygen and, with greater difficulty, carbon will be possible at resolution about 20 times worse than that of protons. With present technology, this relationship corresponds to spatial resolution on the order of 6 cm. It is the author's opinion that results obtained at this level of resolution may not justify the expense and difficulty of achieving them.

With present capability, sodium or phosphorus NMR whole-body imaging is possible and is on the verge of becoming practical. The emergence of such imaging awaits only suitable problems to which it can be applied beneficially, in view of the limited spatial resolution each element will afford.

For the future, the potential of NMR imaging of other nuclei depends on two variables. The first variable is improvement of the technology. For example, if 1-mm^3 resolution becomes routine for proton NMR imaging, such nuclei as potassium, chlorine, oxygen, carbon, and even deuterium may then be capable of yielding images with resolution between 1.5 and 2.5 cm. This level of resolution is almost as good as in photon-emission tomography images. The second variable is applicability. As conventional NMR study of these other nuclei continues, specific applications may arise for which NMR imaging would be useful. Should both variables become more favorable, NMR imaging of nuclei other than protons will become clinically relevant.

REFERENCES

1. Lauterbur PC: Spatially-resolved studies of whole tissues, organs and organisms by NMR zeugmatography, in Dwek RA et al (eds): *NMR in Biology*. London, Academic Press, 1977, pp 323–335.
2. Crooks L, Hoenninger J, Mitsuaki A, et al: Tomography of hydrogen with nuclear magnetic resonance and the potential for imaging other body constituents. *SPIE J* 206:120–128, 1979.
3. Bowen HJM: *Trace Elements in Biochemistry*. London and New York, Academic Press, 1966.
4. Hoult DI, Richards RE: The signal-to-noise ratio of the NMR experiment. *J Magn Reson* 24:71–85, 1976.
5. Hoult DI, Lauterbur PC: The sensitivity of the zeugmatography experiment involving human samples. *J Magn Reson* 34:425–433, 1979.
6. Bottomley PA, Andrew ER: RF magnetic field penetration, phase shift, and power dissipation in biological tissue: Implications for NMR imaging. *Phys Med Biol* 23:309–317, 1978.
7. Gadian DG, Robinson FNH: Radio frequency losses in NMR experiments on electrically conducting samples. *J Magn Reson* 34:449–455, 1979.
8. Libove JM, Singer JR: Resolution and signal-to-noise relationships in NMR imaging in the human body. *J Phys E* 13:38–44, 1980.
9. Censor Y, Gustafson DE, Lent A, et al: A new approach to the emission computerized tomography problem: Simultaneous calculation of attenuation and activity coefficients. *IEEE Trans Nucl Sci* NS-26:2775–2779, 1979.
10. Chang L-T: Attenuation correction and incomplete projection in single photon emission computed tomography. *IEEE Trans Nucl Sci* NS-26:2780–2789, 1979.

11. Budinger TF: Thresholds for physiological effects due to RF and magnetic fields used in NMR imaging. *IEEE Trans Nucl Sci* NS-26:2821–2825, 1979.
12. Llinas M, Wüthrich K, Schwotzer W, et al: ^{15}N NMR of living cells. *Nature* (*London*) 257:817–818, 1975.
13. Wüthrich K: *NMR in Biological Research: Peptides and Proteins*. Amsterdam and New York, North-Holland and American Elsevier, 1976, pp 293–316.
14. Burt CT, Glonek T, Bárány M: Analysis of living tissue by phosphorus-31 magnetic resonance. *Science* 195:145–149, 1977.
15. Dawson MJ, Gadian DG, Wilkie DR: Contraction and recovery of living muscles studied by ^{31}P nuclear magnetic resonance. *J Physiol* (*London*) 267:703–735, 1977.
16. Hollis DP, Nunnally RL, Taylor GT, et al: Phosphorus nuclear magnetic resonance studies of heart physiology. *J Magn Reson* 29:319–330, 1978.
17. McLaughlin AC, Takeda H, Chance B: Rapid ATP assays in perfused mouse liver by ^{31}P NMR. *Proc Nat Acad Sci USA* 76:5445–5449, 1979.
18. Chance B, Radda G, Seeley PJ, et al: ^{31}P NMR of excised and *in situ* brain tissues, in *NMR and Biochemistry*. New York, Marcel Dekker, 1979, pp 269–281.
19. Ackerman JJH, Bore PJ, Gadian DG, et al: NMR studies of metabolism in perfused organs. *Philos Trans R Soc London Ser B* 289:425–436, 1980.
20. Grove TH, Ackerman JJH, Radda GK, et al: Analysis of rat heart *in vivo* by phosphorus NMR. *Proc Nat Acad Sci USA* 77:299–302, 1980.
21. Ackerman JJH, Grove TH, Wong GC, et al: Mapping of metabolites in whole animals by ^{31}P NMR using surface coils. *Nature* (*London*) 283:167–170, 1980.
22. Chance B, Nakase Y, Bond M, et al: Detection of ^{31}P nuclear magnetic resonance signals in brain by *in vivo* and freeze-trapped assays. *Proc. Nat Acad Sci USA* 75:4925–4929, 1978.
23. Bendel P, Lai C-M, Lauterbur PC: ^{31}P spectroscopic zeugmatography of phosphorus metabolites. *J Magn Reson* 38:343–356, 1980.
24. Lauterbur PC, Kramer DM, House WV Jr, et al: Zeugmatographic high resolution nuclear magnetic resonance spectroscopy. Images of chemical inhomogeneity within macroscopic objects. *J Am Chem Soc* 97:6866–6868, 1975.
25. Brooks RA, Di Chiro G: Theory of image reconstruction in computed tomography. *Radiology* 117:561–572, 1975.
26. Hoult DI: Rotating frame zeugmatography. *J Magn Reson* 33:183–197, 1979.
27. Kumar A, Welti D, Ernst RR: NMR Fourier zeugmatography. *J Magn Reson* 18:69–83, 1975.
28. Brown T, Ugurbil K, Kincaid BM: Paper Presented at 21st Experimental NMR Conference, Tallahassee, Florida, March 1980.
29. Bottomley PA: Personal communications, 1980.
30. Hinshaw WS: Image formation by nuclear magnetic resonance: The sensitive point method. *J Appl Phys* 47:3709–3721, 1976.
31. Damadian R, Goldsmith M, Minkoff L: NMR in cancer: XVI. FONAR image of the live human body. *Physiol Chem Phys* 9:101–104, 1977.
32. Crooks LE, Grover TP, Kaufman L, et al: Tomographic imaging with nuclear magnetic resonance. *Invest Radiol* 13:63–66, 1978.
33. Gordon R, Hanley P, Shaw D, et al: Paper Presented at 21st Experimental NMR Conference, Tallahassee, Florida, March 1980.

34. Localisation of metabolites in animal and human tissues using ^{31}P topical magnetic resonance, Oxford Research Systems Product and Application Note, Ferry Hinksey Road, Oxford OX2 ODT, 1980.

35. Delayre JL, Ingwall JS, Malloy C, et al: Sodium-23 nuclear magnetic resonance images of an isolated perfused working rat heart. *Science* 212:235–236, 1981.

36. Brooker HR, Hinshaw WS: Thin-section NMR imaging. *J Magn Reson* 30:129–131, 1978.

37a. Shporer M, Zemel H, Luz Z: Kinetics of complexation of sodium ions with valinomycin in methanol by ^{23}Na NMR spectroscopy. *FEBS Lett* 40:357–360, 1974.

37b. Ostroy F, James TL, Noggle JH, et al: Studies on the characterization of the sodium-potassium transport adenosinetriphosphatase. NMR studies of ^{23}Na binding to the purified and partially purified enzyme. *Arch Biochem Biophys* 162:421–425, 1974.

38. Goldberg M, Gilboa H: Sodium magnetic resonance in biological systems. Interpretation of the relaxation curves, in *Nuclear Magnetic Resonance Spectroscopy in Molecular Biology*. Dordrecht, The Netherlands, D Reidel Publishing, 1978, pp 481–491.

39. Monoi H: Nuclear magnetic resonance of tissue ^{23}Na. *Biophys J* 14:645–651, 1974.

40. Spector WS: *Handbook of Biological Data*. Philadelphia, WB Saunders, 1956.

41. Wüthrich K: *NMR in Biological Research: Peptides and Proteins*. Amsterdam and New York, North-Holland and American Elsevier, 1976, pp 329–334.

42. Chmelnick AMA, Fiat D: Applications of ^{17}O magnetic resonance in paramagnetic systems, in Axenrod T, Webb GA (eds): *Nuclear Magnetic Resonance Spectroscopy of Nuclei Other Than Protons*. New York, John Wiley & Sons, 1974, pp 143–152.

43. Allerhand A, Dill K, Goux WJ: Applications of natural abundance carbon-13 NMR to studies of proteins and glycoproteins, in *NMR and Biochemistry*. New York, Marcel Dekker, 1979, pp 31–50.

44. Holland GN, Bottomley PA, Hinshaw WS: ^{19}F magnetic resonance imaging. *J Magn Reson* 28:133–136, 1977.

45. Andrew ER, Bottomley PA, Hinshaw WS, et al: NMR images by the multiple sensitive point method: Application to larger biological systems. *Phys Med Biol* 22:971–974, 1977.

46. Cox SFJ, Styles P: Towards biochemical imaging. *J Magn Reson* 40:209–212, 1980.

47. Hoffman RA, Forsen S: High resolution nuclear magnetic double and multiple resonance, in Emsley JW, Feeney J, Sutcliffe LH (eds.): *Progress in NMR Spectroscopy*. Oxford, Pergamon, 1966, vol 1, pp 15–204.

48. Lauterbur PC, Dias MHM, Rudin AM: Augmentation of tissue water proton spin-lattice relaxation rates by *in vivo* addition of paramagnetic ions, in *Frontiers of Biological Energetics*. New York, Academic Press, 1978, vol 1, pp 752–759.

49. Hollis DP, Bulkley BA, Nunnally RL, et al: Effect of manganese ion on the phosphorous NMR spectra of the perfused rabbit heart: A possible new membrane probe. *Clin Res* 4:240A, 1978.

IV

Potential Hazards

10

Potential Medical Effects and Hazards of Human NMR Studies

Thomas F. Budinger, M.D., Ph.D.

Two general categories of in vivo NMR studies are high-resolution NMR imaging of proton density and relaxation parameter distribution at moderate magnetic fields (0.1-0.2 Tesla), and in vivo NMR sampling at high magnetic fields but low resolution for delineation of phosphorus energy states. Other NMR imaging applications, such as distribution studies of ^{19}F and ^{23}Na or other elements, fall between these two extremes. The potential sources of detrimental health effects of patients exposed to in vivo NMR techniques can be conveniently placed into three types:

1. Effects of static magnetic fields up to 2 Tesla (20,000 G)
2. Effects of induced currents associated with rapidly varying magnetic fields used for spatial localization in some instruments
3. Heating effects from magnetic fields associated with the radio frequency coil used in the NMR imaging apparatus.

The NMR health-hazards problem includes two categories of physical phenomena that have long been a subject of controversy. The first category involves the question of effects from high steady magnetic fields. The second category involves the effects of electromagnetic fields on biologic systems. These effects have received emphasis during and after World War II, when heating effects from microwaves became a health problem for personnel around radar installations. In addition, reports from the USSR regarding effects at very low levels of incident power density of electromagnetic radiations fueled conjectures that there were important effects other than heating associated with the induced electric field from penetrating radio frequency radiations. Relevant reviews of old as well as new concepts in these two areas can be found in books, articles (1-3), and reports of recent symposia (4-6). Induced electric-current densities and their potential for biologic hazards have received relatively little attention, however.

This chapter will attempt to place in perspective present knowledge about potential hazards and to suggest guidelines, thresholds, and procedural approaches that might minimize risk and, at the same time, not stifle progress in NMR physiologic and clinical studies. The goal of this chapter is to present the key issues and central physical parameters that are involved in evaluating the thresholds for potential hazards. This chapter will also attempt to update previous discussions of potential hazards of NMR imaging (7-9). Because we now have a better understanding of the mechanisms involved in reported observations from static and changing magnetic fields, guidelines suggested by the author (7) based on experiments by others must now be changed quantitatively and qualitatively. In addition, the guidelines for radio frequency exposure are modified to be consistent with direct estimates of absorbed power, rather than a translation of calculated incident power density to absorbed power, as was done previously (7). In summary:

1. For static magnetic fields, harmful effects on human beings and reproducible cellular, biochemical, and genetic effects have not yet been observed at fields of less than 2 Tesla (20,000 G).
2. For pulsing magnetic fields, effects depend on waveform symmetry, maximum amplitude,[a] and rate of field change.
3. For radio frequency fields, a practical upper level for absorbed power is 4 W/kg in medically important studies of short duration (less than 10 min). For long-term studies, 1 W/kg is a reasonable level.

Three basic problems that plague most experiments are lack of proper controls (e.g., sham exposures of controls must include animal or cell-culture handling exactly the same as that for exposed subjects), inadequate temperature control (e.g., $\pm$ 0.1°C required in biophysical experiments), and inadequate field strength to elicit unequivocal results (e.g., >20 Tesla is probably required).

STATIC MAGNETIC FIELDS

There is no positive scientific evidence for detrimental human-health effects from static magnetic fields. The exceptions are low-level field effects

[a] 0.4 Tesla/sec maintained for 25 msec causes visual sensations, but 24.0 Tesla/sec for less than 1 μsec will not cause visual phenomena.

on orientation of bacteria and birds (10), and electrocardiographic abnormalities associated with an induced electromotive force due to blood flow discussed below (11).

Conflicting data can be found in reviews by Barnothy (1), Sheppard and Eisenbud (2), Silver and Tobias (3), Phillips et al. (4), and Tenforde (5). In the vast literature on experiments with cell cultures, animals, and human beings, the author has not found an experimental protocol that, when repeated by other investigators, gives similar positive results. A major experimental result reported in 1976 by Malinin et al. (12a) gave strong evidence for morphologic and physiologic transformation of mammalian cells (fibroblasts) in fields of 0.35 Tesla. It was this work, as well as repeatable experiments of T-wave abnormalities on the electrocardiograms of animals exposed to fields greater than 0.3 Tesla, that led to the previous suggestion of a prudent threshold in the range of 0.35 Tesla for the static magnetic field in NMR imaging studies (7). As will be discussed below, the flow-induced electromotive force accounts well for these electrocardiogram signals (11), and this phenomenon is not considered to be of physiologic importance. It has now been ascertained that the experiments reported by Malinin et al. (12a) cannot be replicated under well-controlled conditions unless slight modifications in the method of treating control as compared with exposed subjects are employed, as apparently was the case in the studies reported previously (12b). Extensive studies by Frazier et al. (13) using magnets at the Stanford Linear Accelerator facility showed that the effects noted by Malinin et al. (12a) could not be repeated unless exposed cultures are handled somewhat differently than controls for a number of cell lines. Other studies on tumor cells are also negative (14).

Positive results on the drug-induced behavior of *Spirostomum* (15) have not been repeated in another laboratory. Genetic effects have not been shown in chromosomal studies by Wolff et al. (9) at fields of 0.35 Tesla and in an extensive series of experiments on the plant *Tradescantia* at fields up to 0.8 Tesla (16).

Electrophysical effects (17) on pineal cells in guinea pigs have recently been reported on exposure to fields comparable to that of the Earth's field (0.1 mTesla). These and other (18) earth-field effects have yet to be verified.

Four proposed mechanisms whereby magnetic fields might influence biologic processes or organism behavior are:

1. Macromolecular orientation changes leading to changes in chemical kinetics and membrane permeability
2. Change in enzyme kinetics by quenching superconductivity that might be present in biologic molecular systems

3. Reduction of nerve conduction velocity
4. Superposition of low potentials (induced electromotive force) on natural biopotentials

Thresholds for the first three mechanisms are 1 to 20 Tesla or above, as will be explained below. Measurable flow potentials of a few millivolts are induced by fields above 0.3 Tesla. These potentials have no known physiologic effect.

Magnetic Interaction Energy and Cell-Orientation Effects

The magnetization, or interaction, energy of a field **H** with biologic molecules of diametric susceptibility is given as:

$$W = -\frac{1}{2}\int_V \mathbf{H}\cdot\overline{\overline{\chi}}\cdot\mathbf{H}\, \mathrm{d}V, \qquad [1]$$

where $\overline{\overline{\chi}}$ is the magnetic susceptibility tensor, and the integration is carried out over the volume of the biologic structure. For fields up to 2 Tesla, the magnetization energy for small biologic subcellular and cellular structure is of the order of 2×10^{-12} erg. This energy is 200 times greater than the room-temperature thermal energy of 4×10^{-14} erg. Thus, compared to the "kT" thermal energy, the magnetization energy is large for 2T. However, suppose the molecule or organized cellular structure has magnetic susceptibility that varies with direction, as is the case with retinal rods (19,20), DNA, liquid crystals, and sickle red cells (21). If this anisotropy in susceptibility is sufficiently large, a turning torque can be expected. The torque, or force, will be proportional to the orientation and difference of the susceptibilities and the magnetic field squared. If n molecules or subcellular units are aligned, the torque will be proportional to the number of units connected in a cooperative fashion. With a sufficiently large anisotropy of each unit and a sufficient number of units, large turning torques can be expected at a field strength below 2 Tesla. Such is the case in repeatable experiments on the orientation effects of retinal rods (19,20) in 1-Tesla fields and the alignment of sickle cells at 0.35 Tesla (21). Even the observations on orientation of bacteria (22) and other animals (10) might be explained by the sensing of the turning torque by nervous tissues. Electron microscope studies (Fig. 10.1) by the author of the number and alignment of single-domain magnetite crystals show an average of 18 magnetite particles of about 400 Å width and length in specimens supplied by Dr. R. Blakemore. The susceptibility difference or magnetic polariza-

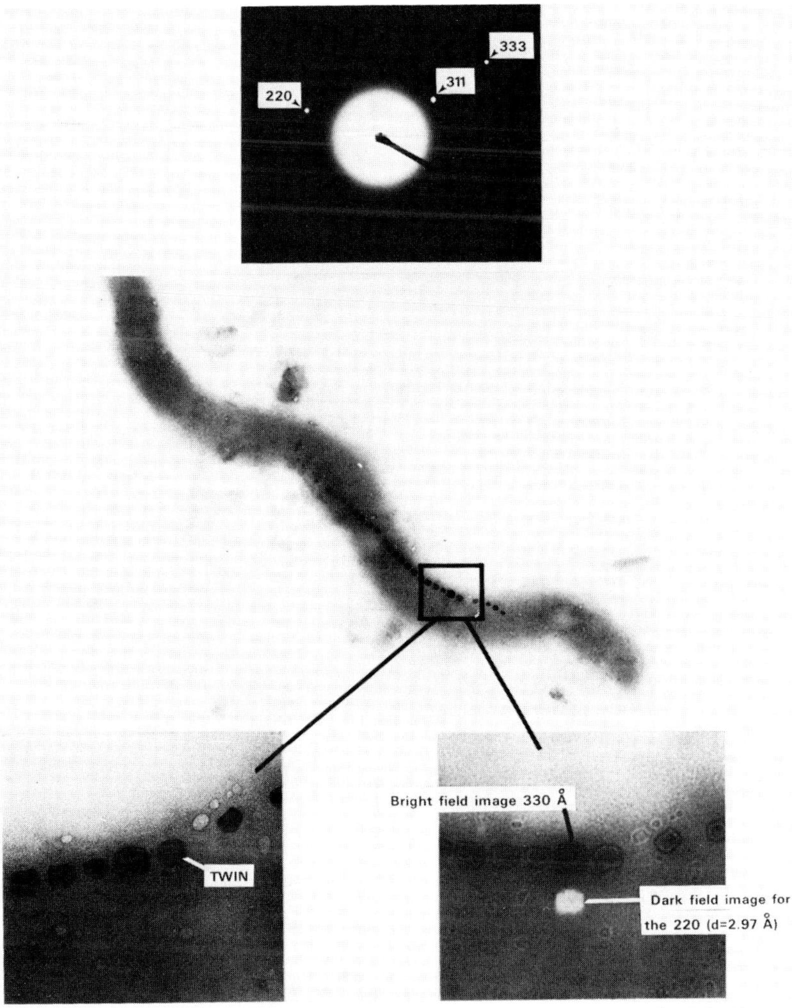

Figure 10.1. Electron micrographs of a freshwater bacterium (supplied by Dr. R. Blakemore). Electron-diffraction studies show that the crystals are single domain, with lattice spacing corresponding to that of magnetite. (Courtesy of T. F. B. Donner Lab., Univ. of Calif.)

tion of these magnetite chains appears ample for an orientation effect at Earth field strengths equivalent to that of sickle cells at 0.35 Tesla.

Effects on Enzyme Kinetics and Quenching of Superconductivity

The presence of superconductivity in organic molecules and its importance to reaction kinetics are controversial issues that involve magnetic field effects because of a well-known effect of magnetic fields on reverting a superconductor to a normal state. The critical magnetic field required to effect this transition might be 20 Tesla. This possibility and the reports of reduction in reaction kinetics at fields of 2 Tesla (23-25) led a group of investigators (26a) to explore a series of enzymatic reactions at fields of 5.0 to 20 Tesla (200,000 G). They found no effects, contrary to previous reports, even though they carefully reproduced the circumstances of others' experiments.

Other recent experiments on the infectivity of DNA after exposure to 2.0 Tesla show no effect (26b).

Effects on Nerve Conduction

The theoretic basis for expecting a change in conduction velocity of nerves is that magnetic fields act on current paths in a way similar to the force on a charge moving in a magnetic field

$$\mathbf{F} = q(\mathbf{v}\times\mathbf{B}), \qquad [2]$$

where q is the charge, $\mathbf{v}$ is velocity, and $\mathbf{B}$ is the magnetic field. This force is a distortion that is perpendicular to both the magnetic field and the original direction of the ion's movement. A magnetic field perpendicular to the current loop pairs associated with the propagation along the nerve will result in a maximal distortion of the current loops and, consequently, a reduction in current velocities. Theoretic examinations of the field strengths necessary for a reduction in action-potential conduction velocities argue that fields required to produce a 10% reduction in conductivity must be 24 Tesla or more (27,28).

Observations that demonstrated large changes in conduction velocity in 10-kG fields (29) and other measurements in a constant inhomogeneous field (30) are probably changes associated with experimental conditions other than magnetic field effects. These experiments could not be duplicated by Schwartz (31) using a lobster-nerve preparation or, more recently, by Gaffey and Tenforde (32) using the frog sciatic nerve. In the latter studies, exposures as long as 4 hr at 2 Tesla did not show a change in the action-potential amplitude, the absolute refractory period, or the conduction velocity. Small changes in submaximal action potentials were found associated with a very slight change in temperature (2% to 4% change in the submaximal action-potential amplitude for a change of 0.1 C).

Electrocardiographic Anomalies

High static magnetic fields cause abnormalities in the electrocardiograms of the heart (33,34). These changes in the T wave at high magnetic fields are related to induced electromotive force potentials due to the flow of blood through the field. Recently, Gaffey et al. (35a) completed a comprehensive study of the T-wave changes in rats, dogs, and baboons. An alteration in the amplitude of the T wave is noted with a threshold near 0.3 Tesla. The T-wave abnormality increases linearly with field to 2T (highest level of the experiments). These observations are consistent with the hypothesis that the electrocardiogram changes are from the induced electromotive force associated with high-velocity blood flow perpendicular to the static magnetic field. The essential features of their observations relevant to NMR imaging are that no arrhythmias or changes in heart rate are induced, the electrocardiogram returns to normal immediately on cessation of field exposure, and blood chemistry of the animals is not altered. What is being observed is an effect quite separate from any physiologic effect and, in fact, Gaffey (35b) has proposed that the waveforms and amplitudes of these induced electromotive forces might be used as a noninvasive method for examining characteristics of blood flow in health and disease. The physical basis of this phenomenon is the well-known voltage-generation phenomenon associated with motion of a conductor in a magnetic field. The induced voltage is derived from the fundamental Equation 2 and is given by

$$\epsilon(\mu\text{volts}) = \frac{B\mathbf{v}d}{100}, \qquad [3]$$

where ϵ is in microvolts, B is in gauss (0.1 mTesla), d is the diameter in centimeters of a vessel through which there is a uniform velocity of $\mathbf{v}$ in centimeters per second. A simple conversion of Equation 3 to volumetric flow gives

$$\epsilon(\mu\text{volts}) = \frac{\text{Flow (cm}^3\text{/sec)}B}{25\pi d}. \qquad [4]$$

The magnitude of these potentials can be appreciated from the example below. The maximal flow in a dog aorta is about 250 cm³/sec, and the aorta's diameter is approximately 1.8 cm. At a field of 0.3 Tesla (3000 G) perpendicular to flow, the induced electromotive force is 4 mV, which is four times the height of the T wave. The induced voltage depends on the position of the heart chambers and great vessels relative to the static magnetic field. The potential measured on the surface of the body will be 10 or more times less than the induced potential change.

A second effect that can contribute an electrocardiogram artifact in high fields is Faraday induction, in which instead of the usual equation

$$\epsilon(\text{volts}) = -10^{-8}\frac{dB}{dt}\pi r^2, \quad [5]$$

we write

$$\epsilon(\text{volts}) = -10^{-8}|\mathbf{B}| \times \frac{da}{dt}, \quad [6]$$

where a is the area perpendicular to the field. The chest diameter changes slightly as the heart forces blood through the thoracic vasculature. Suppose a 20-cm diameter chest changes by 2 mm during a 100-msec interval due to the pulse of blood from a heart contraction. This can cause a maximal change in the area perpendicular to the field of 10 cm^2/sec. In a field of 0.3 Tesla, a voltage of 0.3 mV will be induced. At 2.0 Tesla, the induced voltage in the surface electrode loop will be nearly 2 mV. The area change can be expected to oscillate, and a ripple artifact will appear. This effect is sensitive to electrode orientation but not to flow direction relative to the magnetic field, as is the case for the magnetohydrodynamic effect.

INDUCED CURRENTS FROM RAPID MAGNETIC FIELD CHANGES

The varying fields superimposed on the main static field are used to obtain spatial information by several strategies in NMR imaging (36). These time-varying magnetic fields generate internal electric currents by the well-known Faraday induction phenomenon. Rapid changes in magnetic fields of 3 Tesla/sec can induce fields of 3 $\mu A/cm^2$, and if this current density is maintained for a sufficient period in one direction, nonthermal biologic effects can be expected. Consider the arrangement shown in Figure 10.2, in which the subject is exposed to a changing field dB/dt. The induced electromotive force around the loop is

$$\epsilon(\text{volts}) = -10^{-8}\frac{dB}{dt}\pi r^2, \quad [7]$$

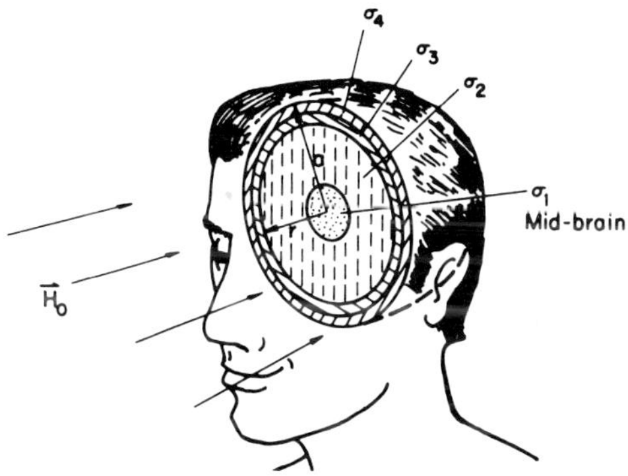

Figure 10.2. Model for calculating induced currents. The σ's refer to various tissue conductivities. Calculations based on this model are merely for order-of-magnitude estimates of the induced current density.

where r is the radius of the loop in which the electromotive force is induced. The E field is given by

$$|\mathbf{E}| = \frac{\epsilon}{\text{loop}} = \frac{\epsilon}{2\pi r}, \qquad [8]$$

and the current density is

$$\mathbf{J} = \mathbf{E}\sigma, \qquad [9]$$

where σ is the tissue conductivity. Thus, the current density is

$$\mathbf{J} = -10^{-8} \frac{dB}{dt} \frac{r\sigma}{2}. \qquad [10]$$

The absorbed power per unit mass of tissue is derived from observing that the energy per time imparted to charged particles by an **E** field is $\mathbf{E}\cdot\mathbf{J}$ (37). Thus, the average absorbed power per unit mass, or the specific absorbed power (SAR) for the induced average electric field $|\mathbf{E}|/\sqrt{2}$, is

$$\text{SAR} = \left(\frac{|\mathbf{E}|^2}{\sqrt{2}}\right)\frac{\sigma}{\rho} = \frac{\sigma}{2\rho}|\mathbf{E}|^2, \qquad [11]$$

where ρ is the tissue density. The energy absorbed is, of course,

$$W = \text{SAR} \times \text{time}. \qquad [12]$$

Using Equation 10, we can estimate the current densities for some assumption of loop diameter. Using Equation 12, we can calculate the energy available to induce biologic effects through electrical phenomena or through heating. Heat production is the main mechanism for biologic effects of the sinusoidal magnetic field changes discussed in a later section.

The threshold current densities required for known biologic effects were evaluated previously using the assumption of a 20-cm diameter loop and the above equations with a conductivity of 0.2 Siemen/m (Siemen = Ohm^{-1}). A working figure of 1 μA/cm for a dB/dT of 1 Tesla/sec was used (7). This working relationship between induced current density and rate of change of magnetic field has two important defects: First, the effective loop size changes by large factors from one experiment to another; secondly, the biologic effect if believed to be strongly dependent on the duration of the electromotive force; that is, some minimal energy is necessary for an effect. Nevertheless, the conversion, 1 $\mu A/cm^2$ for 1 Tesla/sec gives an order of magnitude for the levels of current densities that will be experienced in switching fields used in NMR imaging strategies. The current densities involved in a nerve action potential are 3000 $\mu A/cm^2$; those for induction of heart fibrillation are 100 to 1000 $\mu A/cm^2$. The current density involved in induction of visual light flashes by the application of electrodes across the human head is only 17 $\mu A/cm^2$, as inferred from experiments published in 1947 by Barlow et al. (38) and corroborated by induction of light-flash sensations in partially blind children at frequencies of 10 to 30 Hz (39).

Perhaps the best documented biologic effect of magnetic fields is the induction of visual light flashes by changing magnetic fields. These visual sensations, known as magnetic phosphenes, were first described by d'Arsonval in 1896 (40). Recent important studies of this phenomenon were those done by Barlow et al. in 1947 (41) and the extensive studies done by Swedish investigators, which were published in 1980 (42). According to their data and studies at the Francis Bitter National Magnet Laboratory, the threshold field change for phosphene induction is 2 to 5 Tesla/sec. However, it is important to note that a minimal magnetic field amplitude of approximately 10 mTesla (100 G) is necessary for this effect at a frequency of 20 to 30 Hz (42). In other words, at a dB/dt of 5 Tesla/sec, the duration of current density in a particular direction must be 2 msec. Thus, it is not surprising that in the experiments of Tucker and Schmidt (43), magnetophosphenes were not induced using 60-Hz fields with a maximal amplitude of 5 mTesla. Figure 10.3 shows a composite of experiments by Barlow et al. (41), Lövsund et al. (42), and Tucker and Schmidt (43).

A second area of known biologic effects of low current densities is stimulation of bone healing by induced electric fields using a coil near the fracture site. This technique has recently become an accepted method of therapy. A current density of 10 $\mu A/cm^2$ for each pulse with a pulse repetition rate of about 30 to 60 sec^{-1} can be inferred from studies per-

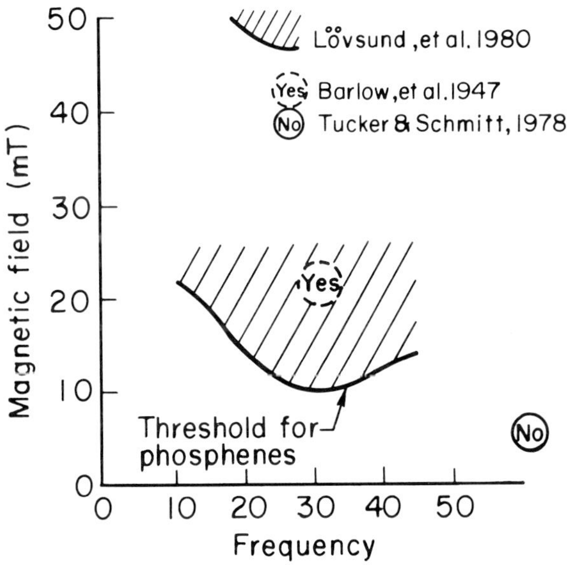

Figure 10.3. Magnetically induced visual sensations (magnetophosphenes) are produced by field changes with the frequency and amplitude thresholds shown.

formed by Bassett et al. (44). Pulses are asymmetric, with a rise time of about 1 μsec. Positive bone healing occurs when these pulses are applied repeatedly over days or weeks. Of relevance to the NMR situation is the fact that sinusoidal waveforms have little or no effect in the range of 30 to 65 Hz, whereas asymmetric waveforms give remarkable changes in cell preparations and in patients.

Two other areas of biologic and medical science effects of induced currents are electroshock therapy and induction of heart fibrillation using capacitive coupling. A review of the electrode configuration and current applied to the head through these electrodes from the work of Fink (45) shows that current densities of 3 mA/cm^2 are applied for 300 msec to cause seizures. Animal experiments have shown that current densities of 1 A/cm^2 cause reversible nerve damage. These current densities are nearly 10^6 times greater than those being used in NMR imaging strategies.

Perhaps the most important area of experimental results with regard to the potential hazards of induced currents is heart fibrillation. Extrapolation of experimental data obtained by Watson et al. (46) from human hearts using 1.8-cm^2 electrodes gives a lower threshold for fibrillating the human heart of 300 μA/cm^2. The minimal fibrillation current density for the dog heart with an electrode area of 0.9 cm^2 is 555 μA/cm^2. Data obtained from a review by Roy (47) and the above observations are sum-

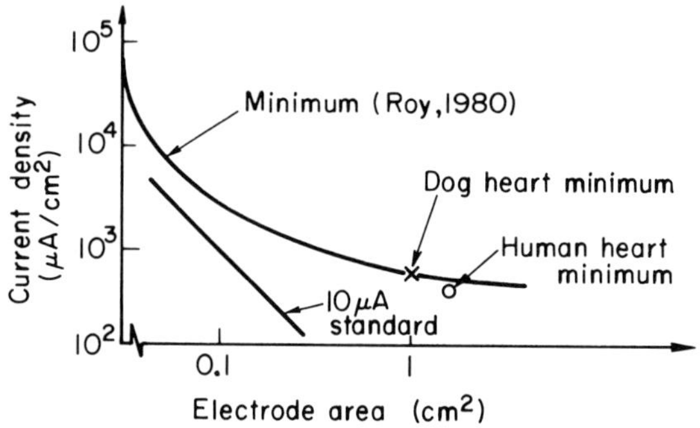

Figure 10.4. Relationship between current density and electrode area for induction of fibrillation in dog and human hearts. The change in field with time for the induction of fibrillation is probably more than 500 Tesla/sec according to these data and the assumption of approximately 1 μÅ/cm^2 for 1-Tesla/sec scaling.

marized in Figure 10.4. Unfortunately, the period during which voltage is applied across the heart did not vary in these experiments. The method of inducing fibrillation involved application of sinusoidal voltages through electrodes of varying size. The state of fibrillation was reached by slowly increasing the voltage. It can be concluded from previous work that the duration of application of a sinusoidal field is 3 sec or less. There was no pathologic evidence of damage at these current densities.

At the Donner Laboratory, we have explored the electrocardiographic behavior of the heart in anesthetized rats exposed for varying periods up to 10 min to field changes of 24 Tesla$\cdot$sec^{-1} at repetition rates of 30 sec^{-1}. The rise time was about 1 μsec. No abnormalities were observed in the electrocardiogram or in recovery from anesthesia.

To investigate the relationship between the magnitude of rapid field changes and biologic effects, we are now conducting experiments in our laboratory to determine:

1. Effects of induced electric current densities on blood-brain barrier permeability
2. Threshold for generating visual sensation (magnetophosphenes)
3. Threshold for platelet activation.

Blood-brain barrier changes noted in the rat for exposures of 1 hr to large field changes (e.g., 10-20 Tesla/sec, with a maximal B field of 6 mTesla) are being conducted using the protocol of Figure 10.5 by Dr. Chi-Kwan Yen of Donner Laboratory. Even though the data were normalized for changes in blood volume of the brain, the wide variability of tracer uptake from one experiment to another suggests that the protocol used is

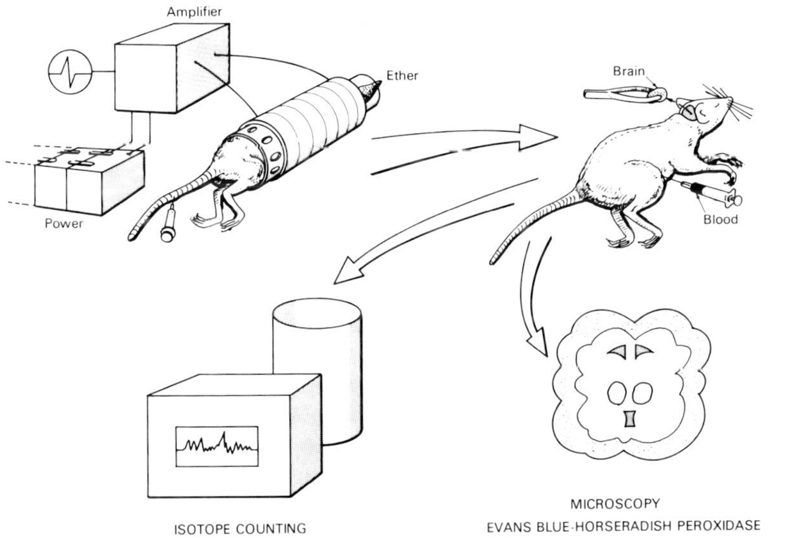

Figure 10.5. Schematic diagram of procedure whereby biologic effects of induced currents are measured in animals and blood samples.

influenced by indeterminate biologic errors (e.g., depth of anesthesia). Thus, statistically valid conclusions cannot be drawn at present.

Recent experimental work done in our laboratory with a maximal dB/dt of 24 Tesla/sec using a coil of 60 turns held close to the eye did not generate phosphenes. The maximum B field was 60 G, and repetition rates of 1/sec, 4/sec, and 30/sec were used. These fields were not sinusoidal, as were those used in experiments shown in Figure 10.3.

Experiments were designed to demonstrate whether platelet changes could be induced by electric current densities associated with 24-Tesla/sec field changes (48). Human platelets were exposed in whole-blood samples immersed in the center of a 200-turn induction coil. The maximal field change was 24 Tesla/sec, with a pulse of 1-μsec rise time and 34-msec decay. The repetition rate for these waveforms was 30/sec, and the exposure times were 30 min. Temperature was controlled at 37°C. After 30 min of exposure, the platelets were separated by ultracentrifugation, embedded, and examined for ultrastructural changes using electron microscopy. Studies on six subjects have been completed. There is a constant overall swelling, with some loss in dense granules in the exposed platelets. Swelling of the internal canalicular system appears to progress to large radiolucent vacuoles (Fig. 10.6). In a hypothesis proposed by Ms. Violet Barghe-Sharghi, who is conducting these experiments, these vacuoles burst at the membrane surface, and membrane rupture results in

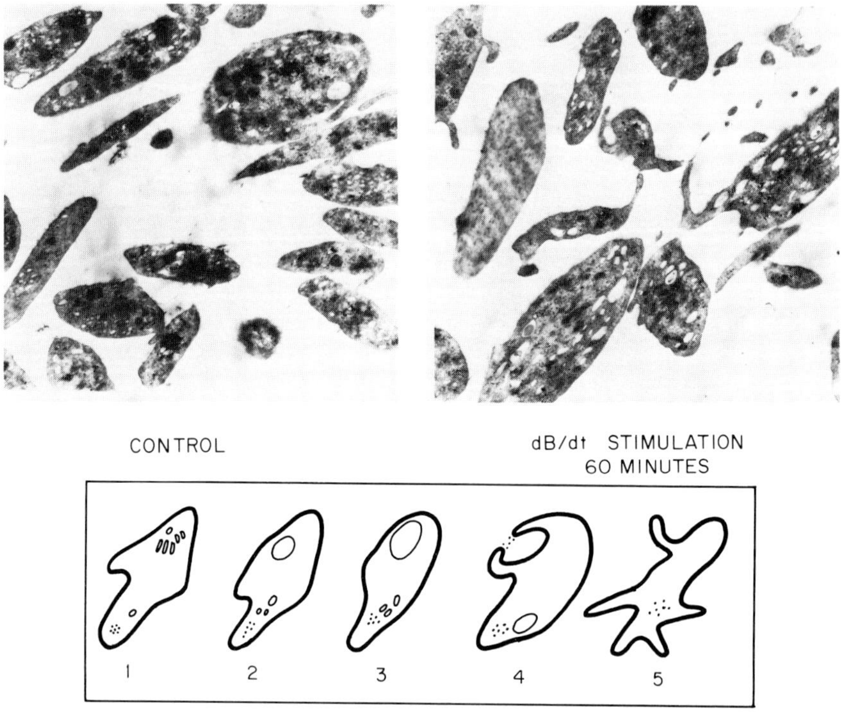

Figure 10.6. Results of 30-min exposure at field changes of 24 Tesla/sec on human platelets (a). The sequence of activation (b) is that suggested by V. Barghe-Sharghi (48).

protuberances that cause a formation of the well-known pseudopodia seen in activated platelets (48).

A plot of the hypothetic relations between the induced current density and time is given for several end-points in Figure 10.7. Clearly, the magnitude of induced current density necessary for induction of heart fibrillation is about 0.1 mA/cm^2, and that required for induction of light flashes is far less. The safe threshold, as has been emphasized previously (7), is a complex function of waveform, amplitude, time duration, and repetition rate. Present evidence is inadequate; thus, a comprehensive guideline is not appropriate at this time. A working level that appears safe is an absorbed energy equivalent to that at each field excursion of a 20-Hz sinusoidal field change with a maximal amplitude of 5 mTesla (50 G). An interim guideline for design of systems with sinusoidally switching fields at frequencies of less than 1 to 200 Hz is to keep the maximum oscillating B field below 50 G.

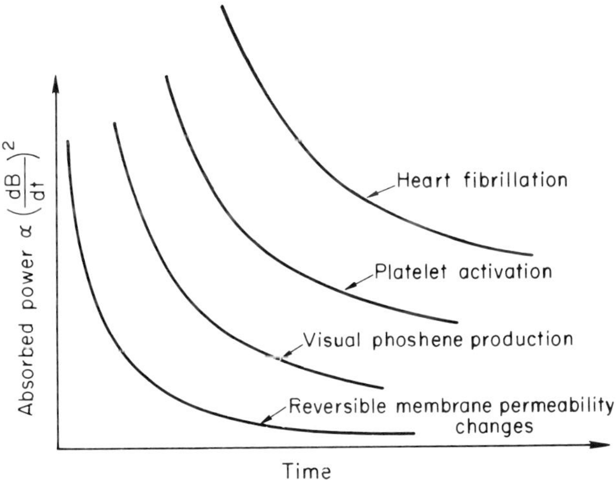

Figure 10.7. Probable relationship between absorbed power or induced currents and the biologic changes suggested. The order of the family of biologic changes has not been confirmed with reference to reversible membrane-permeability changes.

RADIO FREQUENCY MAGNETIC RADIATION EFFECTS

The known radio frequency magnetic field hazards are related to tissue heating. The heating problem from propagating electromagnetic radiation is conventionally approached by calculation of the absorbed energy that results from field exposures to incident electromagnetic power densities (37,49,50). In the past, tissue heating has been investigated from measurements of the incident power density, frequency of the radiation, and size of the object. This approach led to an appreciation of the fact that there is as much as a 100-fold difference between absorbed power in a rat and that in adult human beings for the same frequency and incident power density. Estimations using the incident power density, as ascertained from the surface of the object, and the average power of the radio frequency system are not appropriate for the NMR imaging situation because the radio frequency radiation is magnetic and the object is in the near field. A somewhat better theoretic estimation based on an assumption of a plane wave incident on tissue slabs of infinite extent has also been employed for the NMR imaging situation (8).

Perhaps the best method for estimating the potential heating hazards in NMR studies is to measure the absorbed power using the ratio of the *Q*s[b]

[b]The *Q* of a coil is the ratio of the energy stored to the energy lost due to coil resistance.

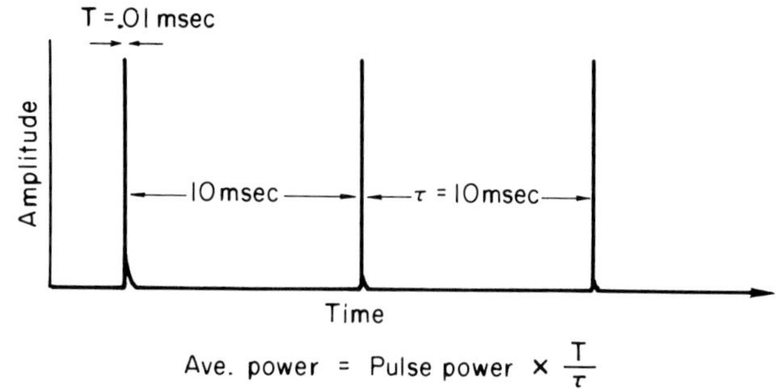

Figure 10.8. Average power is related to peak power, duration of pulse, and pulse repetition rate. Typical sequence from the Nottingham group.

of the radio frequency coil for the loaded to the unloaded conditions. That is,

$$\text{SAR} = \alpha\left(\frac{Q_L}{Q_u} \times P_{\text{max}} \times \frac{T}{\tau}\right)\frac{1}{V_\rho} \text{ (W-kg)},$$

where α is the fraction of absorbed energy not radiated or reflected out of the system, Q_L is the measured Q with the subject in the coil, Q_u is that of the unloaded coil, P is the power of each pulse in watts, T is the time duration of each pulse, $1/\tau$ is the number of pulses per unit time (Fig. 10.8), and $1/V_\rho$ is the mass in kilograms of the object within the effective limits of the coil. The equation with $\alpha = 1$ should give a reasonable measure of the total absorbed power, which might be 50% or more of that which is delivered to the radio frequency coils. The reflectivity at the air/tissue interface might be as high as 60% at the NMR frequencies used. This reflectivity will, of course, decrease the amount of absorbed power and, therefore, heating. Most heating will occur at the body's surface, and, due to the geometry and dielectric constant variations at tissue interfaces, hot spots can be expected to occur (51,52). The skin depth δ, the depth at which the power is reduced by $(1/e)^2$, is shown as a function of frequency for tissues in Table 10.1 (50).

Table 10.1. Skin Depth δ as a Function of Frequency

Frequency (MHz)	δ *(cm)*
1	91
10	22
41	11
100	7

The prudent threshold value for NMR studies is 1.5 W/kg for long-duration studies and 4 W/kg for studies of 10 min or less. The bases for these suggested absorbed power guidelines are given below.

The safety level established by the Occupational Safety and Health Administration is 10 mW/cm^2 incident electromagnetic power density for 10 mHz to 100 GHz averaged over a 6-min period. This standard was similar to that of the American National Standards Institute (ANSI), which is derived from the concept that the induced heat load should not exceed the basal metabolic rate of man, which is 1.5 W/kg during sleep and 15 W/kg for heavy exercise. Recently, this standard has come up for revision primarily because of the results of a five-year investigation by the ANSI subcommittee C95.4, which concluded that biologic effects thresholds from observations reported in the literature are 4 W/kg absorbed power or more. They added a safety factor of 10 and therefore, using a level of 0.4 W/kg, have derived new standards that are based on the frequency dependence of this specific absorbed power in children and adults using the previous calculations by Durney et al. (37). Thus, the new standards[c] not yet officially adopted will give the allowable incident power density for various frequency intervals.

The important conclusion from the work of the ANSI subcommittee is the determination that 4 W/kg of absorbed power is a threshold for biologic effects because this number can be used as a baseline for ascertaining possible heating hazards from the NMR radiofrequency magnetic field to which subjects are exposed under various pulse-timing sequences and power used in NMR imaging. The body has the ability to lose heat by radiation and evaporative (Joule-Thompson) mechanisms. The latter mechanism is most effective in low-humidity environments, in which the body can lose 500 kcal/hr. It is instructive to calculate the temperature rise using a soft-tissue-specific heat of 0.83 kcal/kg/°C (3.5 J/kg/°C). Assuming the subject has no ability to lose heat or transport the absorbed energy to other parts of the body, the expected temperature rise is

$$\Delta T(°\mathrm{C}) = \frac{\mathrm{SAR} \times t}{4186 \times h},$$

where SAR is in watts per kilogram, t is time in seconds, and h is specific heat. For a specific absorbed power of 4 W/kg for 10 min, the expected temperature rise is

$$\Delta T(°\mathrm{C}) = \frac{4\ \mathrm{W/kg} \times 600\ \mathrm{sec}}{4186 \times 0.83\ \mathrm{kcal/kg/°C}}$$
$$\doteq 0.7\ \mathrm{C}.$$

[c]The new standard allows exposures of 100 mW/cm^2 from 300 kHz to 3 MHz, decreasing as the square of the frequency to 1 mW/cm^2 at 30 MHz From 30 to 300 MHz, the proposed standard is 1 mW/cm^2.

Conduction of heat away from the tissue volume will decrease this heat burden; however, most energy will be absorbed at the surface, and hot spots will occur (52). Thus, a conservative guideline for long-term exposures (e.g., 30 min) is the average metabolic waste energy of 1.5 W/kg. For exposures less than 10 min, 4 W/kg should be tolerated at frequencies less than 10 MHz.

The proposed nonthermal effects of radio frequency fields on human behavior and physiology is a controversial subject at present because no mechanism has been proved for the observed effects of weak fields of 100 Hz on animals and human beings (53-55).

Effects of electromagnetic fields on calcium transport that are repeatable and supported by experimental data show a 15% to 20% change in the calcium bound to brain cells after a 20-min exposure at 1 to 75 Hz. It was proposed that the mechanism for these observations was a field-induced alteration in the binding of calcium ions that would lead to conformational changes in nerve tissue membranes and a consequent shift in the threshold for excitation. A recent study by Tenforde (56) has shown that the observed alteration in calcium ion binding does not occur through a thermal energy transfer mechanism. The field interaction may involve a resonant or cooperative energy transfer mechanism. At present, although several physical models have been proposed for nonthermal interactions at low field intensities, there are no biophysical measurements that would tend to support one or the other mechanisms (56).

At high field intensities and frequencies in the range of 1 to 100 MHz, a "nonthermal" effect is characterized by alignment of particles of charcoal, starch, milk, or blood cells to form chains parallel to the electrical lines of force. This chain formation, known as the bead-chain effect, has been quantified, and the proposed mechanism is an attraction between particles in which dipole charges are induced by the radio frequency fields. It appears, however, that this effect will not occur at levels below that at which thermal effects occur (57).

Changes in the blood-brain barrier permeability may be nonthermal effects noted at power densities less than 10 mW/cm^2 in small animals using gigahertz frequencies. It is noteworthy that the specific absorbed power at this frequency is about 5 W/kg and 100 times greater than the SAR for a human being at the same exposure (7,37). This is an experimental area of great controversy. Experiments of Oscar and Hawkins (58) and Albert (59) showed blood-brain barrier permeability changes associated with gigahertz in the rat using fluorescein, mannitol, and horseradish peroxidase in microscopic and analytic studies. These observations could not be repeated in studies by others (60-62).

The most recent study and review by Oscar et al. suggests that previously reported small increases in blood-brain barrier permeability may be a secondary effect caused by microwave alteration of blood flow, blood pressure, or blood vessel area (63). Because these and other experiments with gigahertz frequencies, small animals, and incident power densities of 10 to 15 mW/cm^2 result in high power absorption and possible hot spots, this author cannot discount local heating as the biophysical mechanism for observed effects.

EXPERIENCES OF NMR RESEARCHERS

Over the past few years, enthusiastic researchers in England, Scotland, and the United States have been exposed to fields of 0.1 Tesla for periods of up to 4 hr. No morbid sequelae have been reported in personal communications (see acknowledgments).

The best evidence for no health hazards from very high fields is the in vivo ^{31}P energy metabolism experiments of Chance et al. (64), Radda and Seeley (65), and Nunnally and Bottomley, in which no changes in brain and heart energy metabolism were noted during studies in animals and no late sequelae were seen in the laboratory animals after the experiments.

EPIDEMIOLOGIC PERSPECTIVES

The health effects on human beings can be investigated by prospective epidemiologic studies or retrospective surveys. Well-controlled studies require years of effort on the part of dedicated and experienced researchers. Some small-population studies have been accomplished in the past, but the observations have been for the most part anecdotal.

The first analysis of the experiences of physicists and technicians working in fields up to 2 Tesla in nuclear physics laboratories was a survey by Beischer (67), who concluded that no deleterious effects were experienced, even after cumulative exposures to fields of 0.5 Tesla for 3 days per year per person. The sensations of tooth pain and taste reported in Beischer's study were due to movement of the head through a field gradient that produces an induced electric current effect, as discussed above. The most alarming report on effects of static fields was published 16 years ago by Soviet investigators who presented data from 1500 workers whose hands were exposed to fields of 35 to 350 mTesla (68). In addition to general symptoms of headache and fatigue, some workers had hand swelling and skin desquamation. There were no controls in this study, and no corroborating reports exist, although hand exposures at similar or higher levels have been experienced by NMR researchers, bubble-chamber technicians, and technical support staff at magnet facilities (69).

The basic problem is that no well-designed epidemiologic study has been completed with a sufficiently large population of exposed persons to show any integrative effects of magnetic field exposures. The major problem in studies aimed at showing the presence or absence of effects by evaluating the change in incidence of some phenomenon or disease is the need to study a sufficiently large sample. For example, the incidence of cancer is about 2.5 in 1000 and for cardiovascular disease is 5 in 1000. With an exposed group of 1000 persons, a threefold increase in the incidence of cardiovascular disease would have to have occurred before a firm conclusion (90% confidence) could be made that there is a health effect from magnetic field exposures (70). Studies are under way on workers at high-energy physics laboratories exposed to fields as high as 1 Tesla frequently during their working careers. These studies include 750 exposed and 750 matched controls and will be completed in late 1981.

It is not clear at this time whether a prospective study of patients exposed to the NMR imaging procedure is warranted because of the difficulty in separating radio frequency effects from magnetic field effects, if any, and the fact that we have no hypothesis regarding what effects one might expect. Perhaps, two prudent protocols are warranted:

Protocol 1. Follow a series of at least 300 persons who have undergone head-imaging studies at magnetic fields less than 0.3 Tesla but relatively high radio frequency power exposures. End-points might be sensory organ performance, brain metabolism studies using NMR or emission tomography, and general health variables.

Protocol 2. Follow a series of at least 300 persons who have undergone high-field ($>$1.5 Tesla) in vivo ^{31}P spectrum measurements using the end-points suggested above.

In a rapidly evolving discipline, one must weigh the relevance as well as the likelihood of completeness when undertaking the responsibility for these types of prospective studies.

SUMMARY

Previous suggestions for guidelines in NMR instrumentation development and clinical studies should be revised because new information has become available and previous experiments have been reanalyzed. Guidelines for magnetic field exposure in industrial activities are necessarily based on different objectives and requirements (71).

Static magnetic fields. Harmful physiologic or genetic effects are not expected at fields of less than 2 Tesla.

Changing magnetic fields. The threshold for effects of induced currents is above that produced from less than 1 to 100 Hz sinusoidal field changes with a maximal field of 5 mTesla (50 G). Waveform, repetition rate, maximal B field, and duration of exposure are variables that require further study.

Radio frequency heating. A practical upper level for absorbed power is 4 W/kg in medically important studies of short duration (less than 10 min). For long-term studies, 1 W/kg is a reasonable level. The power absorbed by the subject can be estimated by measurement of the radio frequency coil Q before and after the subject is placed in the NMR instrument. Large metal objects will absorb power in proportion to the conductivity of the prosthesis.

ACKNOWLEDGMENTS

Ms. Linda Lutgens prepared and edited the manuscript. The author benefited from discussion with Drs. Carl Durney, Jerome Singer, Paul Lauterbur, Waldo Hinshaw, P. Bottomley, Peter Mansfield, Neil Moore, George Holland, George Radda, Britton Chance, Denis Mahlum, Tom Tenforde, Cornelius Tobias, Cornelius Gaffey, Joseph Cobb, George Mullholland, Stephen Derenzo, Leon Kaufman, William Edelstein, and Chi-Kwan Yen.

REFERENCES

1. Barnothy MD (ed): *Biological Effects of Magnetic Fields.* New York, Plenum Press, 1964, 1969, vols 1,2.
2. Sheppard AR, Eisenbud M: *Biological Effects of Electric and Magnetic Fields of Extremely Low Frequency.* New York, New York University Press, 1977.
3. Silver IL, Tobias CA: Magnetic fields and their biological effects, in Tobias CA, Todd PW (eds): *Space Radiation Biology and Related Topics.* New York, Academic Press, 1974, pp 258–292.
4. Phillips RD, Gillis MF, Kaune WT, et al (eds): *Biological Effects of Extremely Low Frequency Electromagnetic Fields.* Proceedings of the 18th Annual Hanford Life Sciences Symposium, Richland, Washington, October 1978. Technical Information Center, Department of Energy, DOE Symposium Series 50, CONF-781016, 1979.
5. Tenforde TS (ed): *Magnetic Field Effects on Biological Systems.* New York, Plenum Press, 1979.

6. Biological Effects of Electromagnetic Waves. Supplement to *Radio Sci* 14(6S), 1979.
7. Budinger TF: Thresholds for physiological effects due to rf and magnetic fields used in NMR imaging. *IEEE Trans Nucl Sci* NS-26(2):2821–2825, 1979.
8. Bottomley PA, Andrew ER: Rf magnetic field penetration, phase shift and power dissipation in biological tissue: Implications for NMR imaging. *Phys Med Biol* 23:630–643, 1978.
9. Wolff S, Crooks LE, Brown P, et al: Tests for DNA and chromosomal damage induced by nuclear magnetic resonance imaging. *Radiology* 136:707–710, 1980.
10. Moore BR: Is the homing pigeon's map geomagnetic? *Nature (London)* 285:69–70, 1980.
11. Gaffey C, Tenforde TF: Alterations in the rat electrocardiogram induced by stationary magnetic fields. Lawrence Berkeley Laboratory, Berkeley, California. LBL-12099, 1981.
12a. Malinin GI, Gregory WD, Morelli L, et al: Evidence of morphological and physiological transformation of mammalian cells by strong magnetic fields. *Science* 194:844–846, 1976.
12b. Mahlum, D: Personal communication, 1981.
13. Frazier ME, Andrews TK, Thompson BB: In vitro evaluation of biomagnetic effects, in Phillips RD, Gillis MF, Kaune WT, et al (eds): *Biological Effects of Extremely Low Frequency Electromagnetic Fields*. Technical Information Center, Department of Energy, DOE Symposium Series 50, CONF-781016, 1979, pp 417–435.
14. Chandra S, Stefani S: Effect of constant and alternating magnetic fields on tumor cells in vitro and in vivo, in Phillips RD, Gillis MF, Kaune WT, et al (eds): *Biological Effects of Extremely Low Frequency Electromagnetic Fields*. Technical Information Center, Department of Energy, DOE Symposium Series 50, CONF-781016, 1979, pp 436–446.
15. Ettienne E, Ripamonti A, Frankel RB: Effect of magnetic fields on the drug-induced contractility of the ciliate Spirostomum, in Tenforde TS (ed): *Magnetic Field Effects on Biological Systems*. New York, Plenum Press, 1979, pp 54–56.
16. Baum JW, Schairer LA, Lindahl KL: Tests in the plant *Tradescantia* for mutagenic effects of strong magnetic fields, in Tenforde TS (ed): *Magnetic Field Effects on Biological Systems*. New York, Plenum Press, 1979, pp 22–24.
17. Semm P, Schneider T, Vollrath L: Effects of an Earth-strength magnetic field on electrical activity of pineal cells. *Nature (London)* 288:607–608, 1980.
18. Baker RB: Goal orientation by blindfolded humans after long-distance displacement: Possible involvement of a magnetic sense. *Science* 210:555–557, 1980.
19. Hong FT, Mauzerall D, Mauro A: Magnetic anisotropy and the orientation of retinal rods in a homogeneous magnetic field. *Proc Nat Acad Sci USA* 68(6):1283–1285, 1971.
20. Geacintov NE, van Nostrand F, Becker JF, et al: Magnetic field induced orientation of photosynthetic systems. *Biochim Biophys Acta* 267:65–79, 1972.
21. Murayama M: Orientation of sickled erythrocytes in a magnetic field. *Nature (London)* 206:420–422, 1965.
22. Blakemore R: Magnetotactic bacteria. *Science* 190:377–379, 1975.
23. Akoyunoglou G: Effect of a magnetic field on carboxydismutase. *Nature (London)* 202:452–454, 1964.
24. Fardon JC, Poydock ME Sr, Basulto G: Effect of magnetic fields on the respiration of malignant, embryonic and adult tissue. *Nature (London)* 211:433, 1966.

25. Levengood WC: Cytogenetic variations induced with a magnetic probe. *Nature (London)* 209:1009–1013, 1966.

26a. Raninovitch B, Maling JE, Weissbluth M: Enzyme-substrate reactions in very high magnetic fields. I. *Biophys J* 7:187–204, 1967.

26b. Tenforde TS: Personal communication, 1981.

27. Liboff RL: A biomagnetic hypothesis. *Biophys J* 5:845–853, 1965.

28. Wikswo JP Jr, Barach JP: An estimate of the steady magnetic field strength required to influence nerve conduction. *IEEE Trans Biomed Eng* BME-*27*(12):722–723, 1980.

29. Reno VR: Conduction velocity in nerve exposed to a high magnetic field. NASA Report No. NAMI-1089, 1969.

30. Kolta P: Strong and permanent interaction between peripheral nerve and a constant inhomogeneous magnetic field. *Acta Physiol Acad Sci Hung* 43:89–94, 1973.

31. Schwartz JL: Influence of a constant magnetic field on nervous tissue. I. Nerve conduction velocity studies. *IEEE Trans Biomed Eng* BME-25(5):467–473, 1978.

32. Gaffey CT, Tenforde TS: Electrical properties of conducting frog sciatic nerves exposed to high DC magnetic fields. *Bioelectromagnetics* 1:208, 1980.

33. Beischer DE, Knepton JC: Influence of strong magnetic fields on the electrocardiogram of squirrel monkeys *Saimiri sciureus). Aerospace Med* 35:939–944, 1964.

34. Togawa T, Okai O, Oshima M: Observation of blood flow EMF in externally applied strong magnetic field by surface electrodes. *Med Biol Eng* 5:169–170, 1967.

35a. Gaffey CT, Tenforde TS, Dean EE: Alterations in the electrocardiograms of baboons exposed to DC magnetic fields. *Bioelectromagnetics* 1:209, 1980.

35b. Gaffey CT: Personal communication, 1981.

36. Hinshaw WS: Image formation by nuclear magnetic resonance: The sensitive point method. *J Appl Phys* 47:3709–3721, 1976.

37. Durney CH, Johnson CC, Barber PW, et al: *Radiofrequency Radiation Dosimetry Handbook*, ed 2. Salt Lake City, University of Utah, 1978, Report SAM-TR-78-22.

38. Barlow HB, Kohn HI, Walsh EG: The effect of dark adaptation and of light upon the electric threshold of the human eye. *J Physiol (London)* 148:376–381, 1947.

39. Budinger TF: Indirect electrical stimulation of the visual apparatus. Lawrence Radiation Laboratory Report UCRL-18347, 1968.

40. d'Arsonval MA: Dispositifs pour la mesure des courants alternatifs a toutes frequences. *C R Soc Bio*. 3:451, 1896.

41. Barlow HB, Kohn HI, Walsh EG: Visual sensations aroused by magnetic fields. *Am J Physiol* 148:372–375, 1947.

42. Lövsund P, Nilsson SEG, Reuter T, et al: Magnetophosphenes; a quantitative analysis of thresholds. *Med Biol Eng Comput* 1981, In press.

43. Tucker RD, Schmidt OH: Tests for human perception of 60 Hz moderate strength low frequency magnetic fields. Ph.D. Thesis, University of Minnesota, 1976.

44. Bassett CAL, Pilla AA, Pawluk RJ: A non-operative salvage of surgically-resistant pseudarthroses and non-unions by pulsing electromagnetic fields. *Clin Orthopaed* 124:128–143, 1977.

45. Fink M: *Convulsive Therapy: Theory and Practice*. New York, Raven Press, 1979.

46. Watson AB, Wright JS, Loughman J: Electrical thresholds for ventricular fibrillation in man. *Med J Aust* 1:1179–1182, 1973.

47. Roy OZ: Technical note: Summary of cardiac fibrillation thresholds for 60 Hz currents and voltages applied directly to the heart. *Med Biol Eng Comput* 18:657–659, 1980.
48. Barghe-Sharghi V, Budinger TF: Human platelet activation by rapid magnetic field changes. Lawrence Berkeley Laboratory, Berkeley, California. LBL-12188, 1981, In press.
49. Schwan HP, Piersol GM: The absorption of electromagnetic energy in body tissues, Part I. *Am J Phys Med* 33:371–404, 1954.
50. Johnson CC, Guy AW: Nonionizing electromagnetic wave effects in biological materials and systems. *Proc IEEE* 60:692–718, 1972.
51. Guy AW: Analyses of electromagnetic fields induced in biological tissues by thermographic studies on equivalent phantom models. *IEEE Trans Microwave Theory Tech* MTT-19:205–214, 1971.
52. Kritikos HN, Schwan HP: Hot spots generated in conducting spheres by electromagnetic waves and biological implications. *IEEE Trans Biomed Eng* BME-19(1):53–58, 1972.
53. Bawin SM, Gavalas-Medici RJ, Adey WR: Effects of modulated very high frequency fields on specific brain rhythms in cats. *Brain Res* 58:365–384, 1973.
54. Adey WR: The influences of impressed electrical fields at EEG frequencies on brain and behavior, in Altshuler H, Burch N (eds): *Behavior and Brain Electrical Activity*. New York, Plenum Press, pp 363–390, 1974.
55. Hamer J: Effects of low level, low frequency electric fields on human reaction time. *Commun Behav Biol A* 2:217–222, 1968.
56. Tenforde TS: Thermal aspects of electromagnetic field interactions with bound calcium ions at the nerve cell surface. *J Theor Biol* 83: 517–521, 1980.
57. Schwan HP: Biophysics of diathermy, in Licht SH, Licht E, (eds): *Therapeutic Heat and Cold*. New Haven, Connecticut, 1965, pp 63–125.
58. Oscar KJ, Hawkins TD: Microwave alteration of the blood-brain barrier system of rats. *Brain Res* 126:281–293, 1977.
59. Albert EN: Light and electron microscopic observations on the blood-brain barrier after microwave radiation, in Hazzard DG (ed): *Proceedings of the Symposium on Biological Effects and Measurements of Radiofrequency/Microwaves*. Rockville, Maryland, Department of Health, Education, and Welfare, 1977, pp 294–304.
60. Merritt JH, Chamness AF, Allen SJ: Studies on blood brain barrier permeability after microwave radiation. *Radiat Environ Biophys* 15:367–377, 1978.
61. Spackman DH, Riley V, Guy AW, et al: Is the blood brain barrier altered by rf irradiation? Paper presented at the URSI Meeting, Seattle, Washington, June 1979.
62. Preston E, Vavasour EJ, Assenheim HM: Permeability of the blood brain barrier to mannitol in the rat following 2450 MHz microwave irradiation. *Brain Res* 174:109–117, 1979.
63. Oscar KJ, Gruenau SP, Folker MT, et al: Local cerebral blood flow after microwave exposure. *Brain Res* 204:220–225, 1981.
64. Chance B, Nakase Y, Bond M, et al: Detection of ^{31}P nuclear magnetic resonance signals in brain by in vivo and freeze-trapped assays. *Proc Nat Acad Sci USA* 75:4925–4929, 1978.
65. Radda GK, Seeley PJ: Recent studies on cellular metabolism by nuclear magnetic resonance. *Annu Rev Physiol* 41:749–769, 1979.

66. Nunnally RJ, Bottomley PA: Assessment of pharmacological treatment of myocardial infarction by phosphorus-31 NMR with surface coils. *Science* 211:177–180, 1980.

67. Beischer DE: Human tolerance to magnetic fields. *Astronautics* 7:24, 1962.

68. Vyalov AM, Shpilberg PI, Ushkevich LB, et al: On the problems of the effect of constant and variable magnetic fields on the human organism, in Shitskova A (ed): *Problems of Occupational Pathology*. Moscow, USSR, Ministry of Health, 1964, pp 169–175.

69. Rubin L: Personal communication, 1977.

70. Budinger TF, Wong PDC, Yen CK: Magnetic field effects on humans: Epidemiological study design, In Phillips RD, Gillis MF, Kaune WT, et al (eds): *Biological Effects of Extremely Low Frequency Electromagnetic Fields*. Technical Information Center, Department of Energy, DOE Symposium Series 50, CONF-781016, pp 379–399.

71. Alpen EL: Magnetic field exposure guidelines, in Tenforde TS (ed): *Magnetic Field Effects on Biological Systems*. New York, Plenum Press, 1979, pp 25–32.

Glossary

Adiabatic fast passage. Excitation of nuclei by fixing either the magnetic field or the radio frequency field and sweeping the other variable, that is, varying the strength of the field through the resonance condition. It is called fast passage because the sweep is fast relative to T_1. Adiabatic fast passage differs from continuous-wave excitation because the net magnetization is not held at a fixed nonequilibrium position; thus, this excitation method can be used to invert spins.

Angular momentum. See *Nuclear angular momentum.*

Boltzmann principle. Principle of thermodynamics that says that when a large number of individual elements is in thermal equilibrium, owing to thermal agitation, individual elements may be excited at all possible energy levels, but the higher energy levels will occur less frequently than will the lower ones. Specifically, the number of nuclei in parallel (N_1) and antiparallel (N_2) alignment is $N_2 = N_1 \exp-(\mu H/kT)$, where μ is the magnetic moment of the nuclei, H is the magnetic field in Gauss, and k is Boltzmann's constant. This small majority of nuclei in the lower-energy distribution is the basis of the resonance phenomenon.

Continuous wave. A technique for achieving NMR. The net magnetic moment is periodically tipped by radiofrequency pulses, the time between which is short compared with the T_1 and T_2 of the sample. The FID from each pulse merges with the spin echo from the previous pulses, forming a continuously observable signal. The net magnetic moment can be conceived of as oscillating about a steady, nonequilibrium position.

Chemical shift. Shielding effect electronic orbital motions have on the magnetic field at a nucleus. The chemical shift is proportional to the applied magnetic field and for protons varies over a range that is 0.001% of the applied magnetic field. Chemical shifts are observable as slightly displaced resonance peaks in high-resolution NMR spectra.

Diamagnetic. A substance in which the internal induced magnetic field is in the opposite direction to, and much weaker than, the external magnetizing field. Aluminum, for example, is diamagnetic.

FID. Free induction decay. If a sample is irradiated with a single 90-degree pulse, the nuclei precess freely, inducing a voltage in the radiofrequency coil that decays as the nuclei lose their transverse magnetization. Thus, theoretically, the signal decays to zero with a T_2 time constant, although it actually decays in a time T_2* that is determined primarily by field inhomogeneities.

Field gradient. A magnetic field whose strength varies along a particular direction.

Fourier transformation. Using Fourier transformations, it is possible to translate a signal that is observable as intensity over time into a signal of intensity versus frequency by resolving the composite sine and cosine waves that constitute the intensity versus time signal. Fourier transformations are used in all imaging techniques. The fast Fourier transformation is a standard computer algorithm.

Gauss. A measure of magnetic flux density. See *Tesla*.

Line width. One of the standard measurements in NMR spectroscopy; the other two are height and frequency. Line width can give an indication of the presence of spin-spin couplings.

Larmor relation. Defines the resonance frequency in a given magnetic field by the formula $f = \gamma H_{\mathrm{Loc}}$, where γ is the magnetogyric ratio and is 42.58 MHz/Tesla for hydrogen.

Lorentzian line. A spectral line with a narrow peak and long tails.

Magnetic dipole. North and south magnetic poles separated by a finite distance. An electric current can create a magnetic dipole moment.

Magnetic moment. Because both individual nucleons and nuclei have a charge distribution, rotation or spin of this charge produces a magnetic dipole whose value is called the magnetic moment. The numeric value of the magnetic moment determines the energies of the different orientations of a nuclei in an external magnetic field.

Magnetogyric ratio. Also known as the gyromagnetic ratio. The magnetogyric ratio is specific for each element, and it can be derived from the equation $\mu/I\hbar$, where μ is the magnetic moment, I is the spin number, and $\hbar$ is Planck's constant over 2π.

Nuclear angular momentum. An intrinsic property of electrons and nuclear particles, associated with their spin. A magnetic moment is associated with the angular momentum of particles.

Nuclear spin number. A variable of the largest measurable component of the nuclear angular momentum (I). Nuclear spin numbers are assigned to all nuclei and are always some multiple of 1/2 or are zero. The number of possible energy states for a given isotope in a fixed magnetic field is $2I + 1$. Any nucleus with spin greater than 0 has a magnetic moment associated with its angular momentum. The spin number is determined by observing the multiplicity and relative spacings in hyperfine structure.

Optimal frequency. Particular range of radio frequencies that will provide the maximal signal-to-noise ratio. The optimal frequency range for NMR imaging still has to be determined, but its upper limit will reflect the decrease of radio frequency penetration at higher frequencies due to the electrical conductivity of the body and the increase in radio frequency coil noise at higher frequencies. The optimal frequency is assumed to be the same for all elements.

Overhauser effect, nuclear. Coupling between two different spin systems. Because it allows saturation transfer or energy transfer between two spin systems, this coupling effect can be used for signal reduction or signal enhancement. It can also be used to study chemical couplings.

Paramagnetic ions. Atoms or ions that slightly increase a magnetic field when placed within it are called paramagnetic. They have an odd number of electrons and a partially filled inner shell (such as is found in transition elements, rare earth elements, and those of the actinide series) and also occur in a few compounds with an even number of electrons (most notably, molecular oxygen) and some metals.

Precessional frequency. Frequency at which nuclei will precess in a given magnetic field. The radio frequency field must match this precessional frequency if it is to induce resonance. In NMR imaging, the precessional frequencies of nuclei fall in the radio frequency range, with wavelengths from 10 to 100 m.

Q factor. Coil-quality factor. A measure of energy loss described as the ratio of energy in the system to energy that is lost in one oscillating cycle.

Quadrupolar nuclei. An electric quadrupolar moment, due to the asymmetric distribution of electrical charge in the nucleus, is found in all nuclei with a spin number greater than or equal to 1. The electric quadrupolar interaction will so broaden the resonance line of elements that sometimes at least half the signal is invisible due to electrical interactions. Molecular movement will create fluctuating electric fields, which will induce transitions among the nuclear quadrupole levels. This nuclear "relaxation" is observed in NMR as if it were induced by a magnetic field. Electric quadrupolar relaxation effects dominate magnetic relaxation effects.

Radio frequency pulse. A short burst of radio frequency radiation. A radio frequency pulse can induce a transverse magnetization that generates a decaying signal (FID). The use of pulsed radio frequency resonance has become standard. With the aid of Fourier transformations in data processing, the frequencies contained in the decaying signals can be distinguished.

Radio frequency bandwidth. Range of frequencies over which energy is observed or generated.

Rotating frame. A coordinate system rotating around the main magnetic field at the same Larmor frequency and in the same direction as the nuclear moments precess. The rotating frame is often used instead of the fixed laboratory frame of reference because in the rotating frame, the magnetization vector describes a much simpler path, just as, on a merry-go-round, passengers seated opposite one another will seem to each other to be remaining still.

Saturation. A condition in which there is no net magnetization. In the case of exposure to a single radio frequency pulse, if $T_2 \ll T_1$, the transverse magnetization disappears before significant repolarization of the spin system occurs, and the sample is said to be saturated.

Signal to noise. In NMR imaging, the nuclear signal increases as frequency squared, whereas coil noise increases as the square root of frequency: The signal-to-noise ratio, ideally, is proportional to $f^{3/2}$. At higher magnetic fields, the signal-to-noise ratio may change because of the increased radio frequency absorption due to the electrical conductivity of the body and of the field dependence of the T_1 and T_2 times of tissues.

Spin. Individual protons and neutrons (nucleons) can be considered as being in orbit around the nucleus, just like electrons. These nucleons also rotate around their axes and thus have a spin like that of electrons. Pairs of neutrons or protons align so that their spins cancel out. A nucleus with an odd number of neutrons, protons, or both will have a net rotational component characterized by an integer or half-integer quantum number called the spin of the nucleus.

Spin coupling. Interaction between the magnetic moment of two adjacent nuclei. Spin-spin coupling can affect the strength of the magnetic field at a nucleus because nuclei that are chemically bonded exhibit an interaction that is preferentially propogated through the chemical bonds due to the interaction between the nuclei and the strongly magnetic electrons in the bond. The extent of spin-spin interactions will determine the transverse relaxation time (T_2) of a tissue. Because the coupling is due to an interaction entirely within the molecule, it is independent of the applied field, H_0.

Spin density. Number of nuclei resonating within a unit volume.

Spin echo. Under certain irradiation conditions, nuclei can be made to resonate so that the axes of their spins point in the same direction at repeated, fixed intervals in time, producing a strong signal called the spin echo.

Spin-lattice relaxation time (T_1). Exponential time constant at which the component of magnetization parallel to the external field grows or decays, that is, reaches equilibrium. This growth or decay results from the interaction of a nucleus with its physical surroundings, hence the name spin-lattice.

Spin-spin relaxation time (T_2). Exponential time constant at which the component of magnetization perpendicular to the external field decays. It results from the interaction of a spinning nucleus with the spin of an identical nucleus pointing in the opposite direction, hence the name spin-spin.

T_1. See *Spin-lattice relaxation time (T_1).*

T_2. See *Spin-spin relaxation time (T_2).*

Tesla. ISU unit of magnetic flux density. One Tesla is equivalent to 10,000 G.

Thermal equilibrium. Heat is evenly distributed between nuclei and their surrounding environment. In a steady magnetic field, thermal equilibrium dictates that more nuclei be in the lower-energy state than in the higher-energy state. Nuclei approach thermal equilibrium with the exponential time constant T_1, which varies in different tissues.

Zeeman effect. Creation of additional energy states due to the application of a magnetic field. The number of fine splittings of the energy levels depends on the number of possible orientations of the angular momentum of the nuclei.

Zeugmatography. Name given by Lauterbur to NMR imaging, from the Greek *zeugma,* to join together, chosen because the NMR image can be regarded as being created by the coupling of the main and radiofrequency magnetic fields by the resonant nuclei.

Index